MUSIC OF THE SUN

MUSIC OF THE SUN
THE STORY OF HELIOSEISMOLOGY

WILLIAM JAMES CHAPLIN

ONEWORLD

OXFORD

MUSIC OF THE SUN

Published by Oneworld Publications 2006

Copyright © William J. Chaplin 2006

ISBN-13: 978–1–85168–451–9
ISBN-10: 1–85168–451–4

Typeset by Jayvee, Trivandrum, India
Cover design by Design Deluxe
Printed and bound in Great Britain by
Biddles Ltd., King's Lynn

Oneworld Publications
185 Banbury Road
Oxford OX2 7AR
England
www.oneworld-publications.com

Learn more about Oneworld. Join our mailing list to
find out about our latest titles and special offers at:

www.oneworld-publications.com/newsletter.htm

In memory of George Richard Isaak

For Alison

CONTENTS

FOREWORD

This book is a historical account of the development of a comparatively young field of science called *helioseismology*, the study of the natural, resonant oscillations of our star, the Sun. The oscillations, which scientists have now been observing for over four decades, are the visible manifestation of standing sound waves trapped in the solar interior. What to the layperson might seem a rather inappropriate, and bizarre, description of the Sun – that of a huge, resonating musical instrument – therefore turns out to be an entirely apposite one.

Observations made over the past half decade or so have also uncovered evidence for resonance in a variety of structures in the solar atmosphere, beyond the visible surface – and by implication beyond the solar interior. The study of these new oscillations has largely been grouped under the heading of 'coronal seismology'. Helioseismology should therefore now be regarded as a field encompassing a resonant choir from the core of the Sun all the way out to the tenuous corona. Here I concentrate solely upon the internal oscillations, and thereby adopt the more traditional definition of helioseismology.

The book I envisaged writing when I foolishly determined to undertake this task was originally to have been rather different. However, as I chatted to and corresponded with my colleagues around the world, and began to accumulate a marvellous repository of recollections, anecdotes and the like, the idea of giving the text a strong, historical, narrative flavour seemed too good to pass up. I therefore have my colleagues to thank for inspiring this change of tack. The words I have committed to the page are of course my responsibility and I hope they will live up to the stories and words of wisdom offered to me.

My gaze has fallen largely on the years when the field was undergoing rapid development in its 'first phase', from beginnings in the early 1960s through to a coming-of-age in the early to mid 1990s. The subsequent rollout of several worldwide networks of observatories dedicated wholly to observations of the oscillations, and the flying of bespoke instruments on the Solar Heliospheric Observatory spacecraft, has seen the field enter a new era. There has also been a natural diversification into several sub-fields.

Although I have endeavoured not only to bring things up to date but also to discuss where we stand in this new era, most of the book is devoted to the first phase, when many of the scientists I introduce in the text made their names. I therefore apologise to present colleagues whom I fail to mention. In five to ten years' time it will be possible to place the achievements of the current era in a clearer, historical perspective in accounts I am sure will follow this one. Limitations of space mean it has not been possible to recount every single development. I have tried to concentrate upon major landmarks and successes and give a flavour of the overall progression of the field. Again, I apologise to those whose work I have omitted and those who receive only a brief mention. The reader will perhaps understand why, to convey something to the non-specialist of how observational research evolves in practice, I have occasionally used examples from my own research group, in Birmingham, to allow me to delve more deeply into how things are done.

I have pitched the level of material at the non-specialist general science reader. Descriptions of the science are therefore non-mathematical and only a few simple equations are used. The text should be useful for university undergraduate students as a wordy resource, which I hope will encourage them to think about the fundamentals of the physics involved. I have also included enough meat, I hope, to satisfy some of the needs of the more specialist readers among my audience.

Some of the footnotes comprise comments of a more technical nature. I have also included references to some of the key papers. This list is not meant to be exhaustive. Rather, it is intended to point those interested in reading further in the right direction.

The book is structured as follows. Chapter 1 is a brief introduction to set the scene. Before I embark on the story of how the field of helioseismology began, Chapters 2 and 3 provide background on the physics at play in a star like the Sun, and the basic principles of resonance. Chapter 2 is a largely chronological account of how knowledge about the Sun built steadily over time, before the advent of helioseismology. In Chapter 3, I introduce the idea of resonance in woodwind instruments and extend these ideas to explain resonant oscillations in stars.

The story of helioseismology begins properly in Chapter 4. Those who like to follow the age-old adage 'begin at the beginning' may want to start here. The non-specialist reader who is more interested in the general story and the personalities involved, or the reader who has good background knowledge of the relevant physics, may prefer that option.

Chapters 4 and 5 tell the story of the first observations of the oscillations, and how, in less than two decades, they provided the means to fully explain the nature of the phenomenon. In Chapter 6, I recount the early theoretical advances, how they went hand in hand with observational developments, and led to the first science successes from helioseismology. Chapter 7 tells how the permanent worldwide networks of observatories and spacecraft instruments which have been dedicated to observing the oscillations were established. Chapters 8, 9 and 10 then bring things more or less up to date on the science side, explaining key successes and their implications, not only for solar and stellar physics but also for areas beyond (like particle physics and cosmology). Finally, Chapter 11 looks at where we stand now, with a brief mention of the immense potential that *asteroseismology* – the extension of observations to other Sun-like stars – holds in store.

Sadly, two of the leading members of the field have passed away in the last few months. I mention George Isaak first because he was someone I was very close to. It was my great privilege to be supervised by and to work closely with George, who died in June. George was one of the pioneers of the field. He was a truly insightful and imaginative thinker, who always, no matter the circumstances, seemed ready with his trademark 'comment and a question'. He was an inspiration, support and a good friend.

John Bahcall passed away only a few weeks ago. He was a prominent leader of the astrophysics community, whose numerous awards and prizes bear testament to the pioneering contributions he made in astrophysics. It was my privilege to have collected personal insights from John, for this book, on his leading work on the solar neutrino problem and the physics of neutrinos.

Bill Chaplin
September 2005
Birmingham

ACKNOWLEDGMENTS

I have many people to thank without whom this book would have been much the poorer in both content and quality. I spent many enjoyable hours talking with colleagues, and these exchanges have proven indispensable to me. I am extremely grateful to those who, further to imparting their wisdom, also read portions of the draft, gave invaluable comments and pointed out errors in the text. (I should add that any errors that remain are entirely my responsibility.) I also express my thanks to those who assisted with photographs and images. My rollcall of gratitude is as follows: Thierry Appourchaux, the late John Bahcall, Sarbani Basu, Tim Bedding, Patrick Boumier, Jørgen Christensen-Dalsgaard, Tom Duvall, Patricia Eliason, Yvonne Elsworth, Eric Fossat, Alan Gabriel, Douglas Gough, Deborah Haber, Steve Hale, Jack Harvey, Frank Hill, Steele Hill, Rachel Howe, David Hughes, the late George Isaak, Umit Isaak, Hans Kjeldsen, Sasha Kosovichev, Don Kurtz, John Leibacher, Ken Libbrecht, Clive McLeod, Brek Miller, Roger New, Bob Noyes, Pere Pallé, Libby Petrick, Ed Rhodes, Teo Roca-Cortés, Phil Scherrer, Takashi Sekii, Clive Speake, Mike Thompson, Steve Tobias, Roger Ulrich, Graham Verner, Sergei Vorontsov and Joe Wolfe. I would particularly like to thank Jørgen, Douglas, Jack and John Leibacher, who each (very generously) spared time to read a sizeable fraction of the entire draft; and Don Kurtz for his helpful feedback.

I wish to thank the anonymous reviewer for their careful report on my first complete draft, which spotted some errors and contained many useful suggestions for improvements; also Martha Jay, Victoria Roddam, Mark Hopwood, Kate Smith, and all at Oneworld; and last, but by no means least, my wife, Alison, for her ever-present support and encouragement, and our cats for their able assistance.

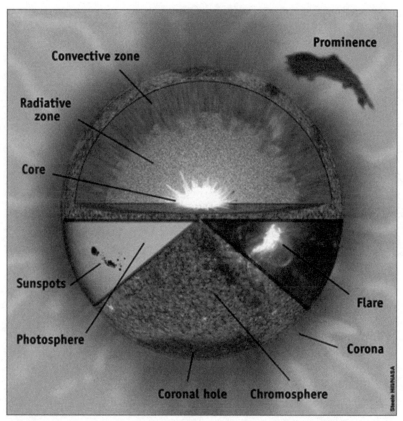

Plate 1. A cutaway showing the interior, and also features at the surface and in the atmosphere above. See the plate section in the middle of this book for a full colour version. Image courtesy of Steele Hill, SOHO, ESA/NASA.

1

INTRODUCTION

It is the early summer of 1960. A physicist and his graduate student are in the process of making groundbreaking observations of the Sun from the Mount Wilson Observatory, overlooking Los Angeles. Their aim is to track honeycomb-like patterns of gas in the solar atmosphere, which, ever restless, continually 'boil' and 'bubble'. In the observations they find something unexpected.

Small patches right across the visible surface appear to be oscillating gently up and down in a clear, repetitive manner. The period of this oscillatory motion is about 5 minutes. The size of excursion each patch undergoes is estimated to be of the order of a few tens of kilometres – but an insignificant fraction of the 700,000-kilometre radius of the star. And soon other observers are able to demonstrate these 'solar oscillations' are present in their data too.

Throughout the 1960s continued observations are made of the phenomenon. The observations get progressively more detailed and knowledge builds. At the same time various theories are advanced to explain the physical nature of the oscillations. Some suggest they may be compressions of gas caused by sound waves in the solar atmosphere. Others say the atmosphere is bobbing up and down like a buoyant cork on water. Then there are others who intimate that matters are not nearly so organised – they say the atmosphere is being buffeted by turbulence.

As the decade draws to a close, two physicists come forward, independently, with a radical idea to explain what is happening. They suggest the key to understanding the oscillations lies not in the atmosphere but rather in the interior of the Sun. The picture they paint casts the Sun in a new light – what if the outer layers of the Sun are resonating like a woodwind instrument?

Sound trapped in a pipe may cause the pipe to resonate at a characteristic set of tones. The new explanation proposes that sound can be trapped in the outer layers of the solar interior, turning the Sun into a huge musical instrument. Though they cannot be 'heard', the Sun's characteristic resonant tones are instead manifested as gentle, rhythmic oscillations of its gas from the compressions of the sound waves. A beautiful pattern of tones is predicted by the theory. But to uncover them requires fresh ingenuity. It takes half a decade to reach this goal. In 1975 a German physicist announces he has observed the characteristic 'notes' of the interior. He just beats a team from California to the discovery.

But the Sun has not given up all its musical secrets. As early as 1972 a physicist from NASA's Goddard Space Flight Centre suggested the whole of the Sun's interior might be pulsating. But even with the discovery of the resonant tones in 1975 the phenomenon is still widely regarded as associated only with the outer layers of the Sun. All that changes in 1979.

Since the early 1970s, teams in Birmingham and Nice have been observing the Sun. The Birmingham group has begun a research programme whose goals are fixed firmly on the distant stars – one being to uncover the presence of planets around other stars. The Sun is at first regarded as a mere stepping stone, a handily located nearby star on which to trial intricate observational techniques. But as the 1970s progress, the focus of the Birmingham team is slowly diverted by the intriguing nature of its solar data. Finally, the team is able to announce a startling discovery – it is the whole of the Sun that is oscillating, not just the outer layers. The resonant tones are truly global. Over the New Year of 1980 the Nice team makes several days of observations from the South Pole which separate the new tones in all their glory. Soon after, a name proposed by a theoretician from Cambridge to describe the new field catches on. The study of the Sun's oscillations is called 'helio-seismology'.

A twenty-year period of discovery – and a new field of science with an intriguing name.

In 1975 the aforementioned scientist from Cambridge, Douglas Gough, was pondering over a suitable name for an important paper he and his graduate student, Jørgen Christensen-Dalsgaard, had just written. The paper discussed the rich insight that observations of many resonant tones of oscillation would give into the interior of our star, the Sun.

The Sun announces its presence to us via the spectrum of electromag-netic radiation it emits, visible sunlight being a small part of this. However, it is not possible to see light from beneath the surface because the interior is opaque to radiation. A quote from Shakespeare is apt:

The heaven's glorious Sun,
That will not be deep-searched with saucy looks.

But sound does not meet with the same degree of obstruction. It can tra-verse the whole interior, from centre to surface, in about an hour. Resonances, and the accompanying oscillations, formed by the sound waves can therefore provide an instantaneous probe of the interior. They give precise information on the physical conditions and so open an unprecedented window onto the exotic vista below the surface of the Sun. Since sound penetrates to the centre of the star, in principle the entire volume is cast open to study by this acoustic probe.

In the absence of these oscillations scientists would be denied the chance to examine directly, and in detail, the structure of the Sun's inte-rior. This was the limited state of affairs before the advent of helio-seismology. A very famous quote from the eminent physicist Sir Arthur

Eddington, who laid many of the foundations of our understanding of stars, speaks to this point. In the introduction to his seminal book on the internal structure of stars,[1] which first appeared in 1926, Eddington writes:

> At first sight it would seem that the deep interior of the sun and stars is less accessible to scientific investigation than any other region of the universe. Our telescopes may probe farther and farther into the depths of space; but how can we ever obtain certain knowledge of that which is hidden beneath substantial barriers? What appliance can pierce through the outer layers of a star and test the conditions within?

As Eddington went on to show in his book – and as we shall discuss in the next chapter – even without the 'certain knowledge' of a direct probe it was nevertheless possible to make predictions on the likely structure found inside stars.[2] Eddington hoped such studies would illuminate understanding of the surface phenomena accessible to observation via the Sun's radiations. But predictions of the internal structure had to rest on certain assumptions, and the detail of them could not be tested directly. The limitations of not having direct access to the solar interior were clear.

Eddington was well aware of the diagnostic potential that resonant pulsations of stars could offer. A year later, he wrote the following in a monograph called *Stars and Atoms*:[3]

> Ordinary stars must be viewed respectfully like objects in glass cases in museums; our fingers are itching to pinch them and test their resilience. Pulsating stars are like those fascinating models in

---

[1] A.S. Eddington, *The Internal Constitution of the Stars* (Cambridge, England: Cambridge University Press, 1926), p. 1.

[2] Eddington followed the introductory quote with the caveat, 'The problem does not appear so hopeless when misleading metaphor is discarded. It is not our task actively to "probe"; we learn what we do by awaiting and interpreting the messages dispatched to us.'

[3] A.S. Eddington, *Stars and Atoms* (Oxford: Oxford University Press, 1927), p. 89.

the Science Museum provided with a button which can be pressed to set the machinery in motion. To be able to see the machinery of a star throbbing with activity is most instructive for the development of our knowledge.

But the types of pulsating stars then observed resonated either in a single tone, or two tones at most. Although these would indeed carry information about a star, the modest amount available limited any inference to the gross, average properties. The Sun turned out to be very different. It sustains many millions of resonant tones. The huge increase in information content thereby implied makes it possible to infer, in great detail, how properties vary with position in the Sun's interior.

With that we return to the paper of Jørgen and Douglas, and its title. It was to these detailed matters of inference that the paper was largely devoted. The paper is noteworthy not only for this content but also because the original title was to have included the name 'helioseismology' in an adjectival form. Douglas was nervous about using the word 'seismology'. He was pondering this while conducting some of the work for the paper on a visit to the National Centre for Atmospheric Research, in Colorado. There, he consulted a Greek dictionary, which gave the translation of *seismos* as 'earthquake'. This was enough to deter Douglas. The paper was submitted with the substitute 'heliological' replacing 'helioseismic'. Only on his return to Cambridge did Douglas discover that the correct translation of *seismos* was the more general 'tremor'. The use of the word in a solar context was therefore entirely apt, and 'helioseismology' was recognised as the name of the field from the early 1980s onwards.[4]

---

[4] The first use of 'helioseismology' in the literature appears to have come from the Crimean solar physicists Severny, Kotov and Tsap (who will feature in Chapters 5 and 6). They used the word in the body of a paper that appeared in 1979 (A.B. Severny, V.A. Kotov and T.T. Tsap, *Soviet Astronomy*, 23, 1979 p. 641). Douglas Gough had by then begun the process of disseminating the name throughout the community. The first use I can find by Douglas himself is in a report he wrote in Nature (D.O. Gough, Nature, 293, 1981, p. 703). The meeting he summarised was hosted by Severny, in the Crimea, in September 1981. The final sub-section of the report is titled 'Helioseismology'.

Just as geoseismologists are able to use natural and human-made earthquakes to probe the interior of our planet, so helioseismologists use the Sun's natural resonant tremors to uncover its internal properties. But what can be learned reaches beyond the realm of understanding of the structure and evolution of the Sun and stars. The Sun is a Rosetta Stone for astrophysics – and helioseismology the dictionary to translate the information. Study of the oscillations has allowed us to test basic physics under the exotic conditions found in stellar interiors and has also led to results that have had great bearing on areas as diverse as fundamental particle physics, cosmology and theories of relativity.

The Sun is just over one hundred times larger, and 300,000 times more massive, than the Earth. It has a mean density not much greater than liquid water. But it is composed of gas, most of it ionised. The vast majority of its constituent atoms – about ninety-two per cent – are hydrogen. Just fewer than eight per cent are helium, with a remaining smattering (amounting to about one-tenth of one per cent) of heavier elements.

Temperatures and pressures are so high in its core that nuclei smash together at great velocity. The nuclear fusion that results makes heavier elements out of lighter ones and releases energy to sustain the star. Nearly all this energy is generated within a region occupying the innermost twenty-five per cent or so by radius. Throughout the bulk of the interior, packets of electromagnetic radiation called photons transport the reservoir of energy outwards. However, conditions in the outer third by radius become so opaque to the radiation that huge, overturning circulation currents must instead carry the load. This convection is no longer effective just beneath the visible surface of the Sun, where the photons once more take over, and the energy is eventually radiated away into space.

Because the interior is opaque any single photon must take a torturous trip, lasting about 100,000 years, to get from the core to finally escape the surface. The Sun therefore keeps its internal structure locked away from direct view via radiation.

Exotic, sub-atomic particles called neutrinos are released by the nuclear fusion reactions in the core. They can readily escape the Sun because they haughtily ignore normal matter and seldom interact with it. This anti-social behaviour is both a benefit and a hindrance to any interested scientist. A benefit because the neutrinos can carry information about the hidden secrets of the core out beyond the surface of the Sun, towards the beckoning scientist on Earth; a hindrance because this elusiveness makes neutrinos very hard to detect, since they must interact within the 'detector' of the scientist if they are to reveal their presence and be counted. Although doing science with these particles is therefore very challenging, neutrino observations have played an important part in furthering our understanding of conditions in the core.

Sound waves do not have to overcome the large obstacles put in the way of the radiation. Furthermore, since they are formed by compressions in the solar gas, they are more amenable to observation than the neutrinos. It turns out the whole interior of the Sun is awash with sound. The characteristics of some of these waves are such that they are selected out by the internal properties of the star, making it resonate like a wind instrument. Just as a pipe confines waves in a cavity, within which the resonant tones of the instrument may be excited and then heard, so the Sun acts as a vast, natural cavity for sound and it sings in the same fashion.

The sounds emanating from a wind instrument come from rhythmic changes in air pressure. These changes take place over time periods of well under a second. The sound waves therefore contain many complete cycles of compression *every* second.

The human ear possesses an audible frequency range from about 15 up to 20,000 cycles per second. This covers just over ten octaves (each given by a doubling of frequency[5]). The tones produced by the wind instruments in an orchestra can cover about seven octaves, from

---

[5] The frequency ratio 2:1 is the *octave*. The ratio 3:2 is the *fifth*, and 4:3 the *fourth*.

the lowest note typically achievable on the contrabassoon – a B-flat at just less than 30 cycles per second – all the way approximately to the C of a tiny piccolo flute at just under 4200 cycles per second.

The sound trapped in the Sun causes it to resonate at much lower frequencies than a typical musical instrument. Its strongest tones are periodic on a timescale of about 5 minutes. This corresponds to a frequency of three-thousandths of a cycle per second – one thousand times lower than that of middle D on the musical scale. The full range of solar tones is expected to span just over four octaves of scale, the strongest covering about an octave.

The intensity of sound is often measured in decibels, a logarithmic scale. Every 10-decibel rise corresponds to about a factor-of-ten increase in the measured number of watts per square metre. Normal conversation produces sound at roughly 60 decibels, a large orchestra at about 100 decibels. The intensity associated with one of the strong solar resonances would be equivalent to roughly 180 decibels were it possible to stand close by. (A level of 160 decibels is sufficient to perforate the eardrum; more severe consequences follow for the human body at higher levels.)

Irrespective of the frequency range, one cannot of course hear the resonant tones of the Sun, from the Earth or an observing spacecraft, because of the intervening vacuum. But one can observe them indirectly. The Sun is made of compressible gas. The compressions of the sound waves trapped inside move the solar gas backwards and forwards – the Sun oscillates or pulsates at periods corresponding to those of its natural resonances. The resonant behaviour can therefore be seen from observations of the oscillations. In practice this is usually done in one of two ways.

In the first method, the speed at which the visible surface is moving is monitored. This is determined from the Doppler shift of light waves received from the Sun. The contributions to the motion of the surface from individual modes are tiny in comparison with the size of the star itself, and occur at speeds of the order of a few kilometres per hour.

Impressive precision is possible in the Doppler shift observations. It is like putting the entire human population in a line – currently about 6400 million souls – then being able to pick out correctly a single individual from among them.

As the gas is compressed, the tiny increase in temperature thereby given raises the intensity of light emitted by the Sun. Measurement of these miniscule variations provides the second method for observing the oscillations. The sizes involved are about one-millionth of the average intensity (a task akin to picking out one person from the population of a major city).

In both cases, the effects of the sound waves are therefore coded into the electromagnetic radiation received from the Sun. The instrumentation of the observer is then used to decode the radiation into a form that can be interpreted in terms of a resonant sound-wave phenomenon.

Musical instruments are made in such a way as to generate specific, complex tones of sound. However, design is not a requirement for resonance to take place (although it may make the results more pleasing to the ear). All objects have a set of natural resonances. To excite these into action one often has to give the object a tap – but when one does, the range and nature of the tones produced provide information on the characteristics of the object itself. The resonances of the Sun are natural, like aeolian tones. Just as the blowing wind can make telegraph wires sing or air rushing through a narrow gap can give a whistle, the buffeting of the turbulent motions of gas just beneath the solar surface make sound, and this sound causes the Sun to resonate. By careful observation of the properties of the many resonant oscillations that are excited, a lid may be opened upon the opaque interior and the structure and physics studied in detail. This is the field of helioseismology.

What follows is largely a historical account of the development of helioseismology. We shall meet the scientists who have played a central role, and tell the story of how insight, inspiration and perspiration in the observational and theoretical areas has produced an impressive list of scientific achievements, achievements that have furthered knowledge not only of the Sun and stars but also of several far-reaching areas of physics and astronomy. We commence our historical narrative in Chapter 4. Before then we begin in the next two chapters by laying some groundwork – starting with the Sun itself.

2

OUR STAR, THE SUN

The first observations of structural features on the surface of the Sun most likely date back some three millennia. These were of sunspots, the small, dark patches that give parts of the visible surface of the Sun, the photosphere, a mottled appearance, and were made by Chinese astronomers. If we go forward to an epoch just over one thousand years ago we encounter the first convincing reference to the tenuous outer atmosphere well beyond the visible surface – the corona. This was revealed to astronomers in Constantinople during a total eclipse of the Sun by the Moon.

I am sure many readers will today be familiar with photographs of eclipses. The strange, hazy finger-like structures that show themselves beyond the dark edge of the intervening Moon reveal the presence of plasma stretching out well above the photosphere. It is because the density of this plasma is so low that we cannot see it under normal circumstances – the bright glare from the visible disc of the Sun simply drowns out the light from the corona.

By the early seventeenth century detailed observations of sunspots were possible by the use of the telescope. But nature can be cruel to the scientist. During the latter part of this century spots largely vanished from the visible disc for a period covering almost three human generations. This must have been confusing indeed. The spots re-established themselves in the early eighteenth century. Unknown to astronomers of the time this was the first indication that the physical topography we see on the solar surface today may not be a typical state.

It was also in the seventeenth century that Sir Isaac Newton demonstrated sunlight could be split into the colours of the rainbow by a glass prism. A closer inspection of the coloured bands, made by William Wollaston almost a century and a half later, revealed a series of fine, dark features. These lines now bear the name of the German physicist Joseph von Fraunhofer, who found them independently fifteen years after Wollaston.

Fraunhofer lines are real features in the spectrum of sunlight. They mark out narrow ranges in wavelength where atoms in the photosphere of the Sun have absorbed, or removed, some of the light from the spectrum.[1] The bands appear dark as a result. The exact locations in wavelength depend on the type of atom involved. Furthermore, the temperature and pressure have to be just right for particular absorbing lines to be present.

After the discovery of the Fraunhofer lines it did not take long for the scientific community to realise that identification of the various lines could be used as a means of determining the chemical composition of the solar atmosphere.

Within the space of two years, in the middle of the nineteenth century, observations were made that probably were the first clear records of two types of dramatic, active features – flares and coronal mass ejections. Flares are striking events associated with a sudden, vast release of energy and local brightening across a range of wavelengths in the corona. Coronal mass ejections are immense strands of plasma that the Sun hiccups into space.

---

[1] Observations of shifts in the wavelengths of these lines are one of the staple methods for recording the solar oscillations.

In 1868, observations of the chromosphere, a layer in the atmosphere above the photosphere and below the corona, revealed something odd in the spectrum of light emitted by its gas – a line that could not be accounted for. These observations – made independently by Pierre Janssen and Sir Joseph Norman Lockyer – had in fact uncovered helium.[2] It may seem odd that an element now believed to account for about one-quarter the mass of all visible matter in the universe should not have been known to scientists before this. However, helium keeps its secrets closely guarded on Earth because it is chemically inert.

Detailed observations of sunspots continued throughout the nineteenth century which demonstrated their number varied systematically over time. Samuel Heinrich Schwabe published the first evidence for this variation in 1843; the eleven-year cycle of sunspot activity now carries his name. Shortly after Schwabe published his results, the sunspot cycle was also linked directly to observations of changes in geomagnetic activity.

It took a further sixty-five years for the real nature of sunspots to be uncovered. In the early twentieth century George Ellery Hale showed the spots marked out locations of intense magnetic fields on the visible surface of the Sun – a few hundred times stronger than a typical refrigerator magnet.

By tracking the spots it was possible for Richard Carrington to show the rotation of the Sun carried them round at different rates, dependent on their latitude. This differential pattern implied surface gases rotated more rapidly at the solar equator than at the poles.

But what could be said about the interior of the Sun before helioseismology? Without the benefit of being able to peer directly inside, it was nevertheless possible to say something about the conditions by bringing to bear some basic physics.

---

[2] Lockyer worked in collaboration with chemist Eduard Frankland, who suggested the name 'helium' for the new element.

The Sun is a huge ball of largely ionised gas. Let us treat it as a gas sphere. We assume it is in a contented state of equilibrium – a precise balance exists between the forces at play inside the Sun. It is held together by gravity. But the mutual forces of attraction between its constituent particles try to collapse the sphere in on itself. So another force is needed to offset, or balance, the effects of gravity, and it is pressure that provides the necessary support.

There is more than one type of pressure. Gas pressure dominates in the Sun and arises from the thermal motions of its particles, which transfer momentum. Similarly, photons, the constituent packets of radiation, give rise to radiation pressure. Magnetic fields were mentioned in our previous discussion of active features in the solar atmosphere. 'Tension' in the field lines gives rise to magnetic pressure. The impact of radiation pressure, and most likely also magnetic pressure, on the basic structure and evolution of the Sun is very small indeed.

As one moves closer to the centre of the Sun the greater weight of overlying material raises the gas pressure. Across any layer of the interior it is therefore the resulting difference in pressure that balances the gravitational force. This negative gradient of pressure – higher at the centre, lower further out – means the layers are 'stratified under gravity'. When an archaeologist talks proudly of the stratification of layers in their trench, they are usually referring to distinct, sudden alterations perhaps from a layer of natural soil to one showing evidence of a building floor. Here, the stratification is marked by a smooth, measured change with depth.

How far can we get from the starting assumption that our star – our ball of gas – is in a state of gravitational equilibrium? We can actually do quite well. We use two differential equations to begin with. One describes the balance of pressure and gravitational forces in any given layer, and is called the equation of hydrostatic equilibrium. The other is called the equation of mass continuity. It describes the fact that mass is distributed smoothly through our self-gravitating gas sphere (the total always adding up to give the same number).

By combining these equations it is possible to arrive at an inequality for the central pressure – that is, an expression that says the pressure must be greater than a certain lower limit. The inequality is given in terms of the total mass and radius of the star. The nineteenth-century physicists had estimates of these numbers at their fingertips, the first from observations of the orbits of the planets, the second from direct observation of the Sun. When the numbers are plugged into the inequality, one arrives at the following startling conclusion: the central pressure must be at least several hundred million times higher than atmospheric sea-level pressure on Earth.[3] The implied conditions are exotic indeed.

Next, we might want to seek a lower-limit estimate of the central temperature. We again assume that contributions from radiation and magnetic pressure may be ignored. This leaves us with gas pressure,[4] which derives not only from the temperature of the matter but also from its density.

The manner in which pressure, temperature *and* density relate to one another is called the equation of state of the gas. We take our gas to be perfect, or ideal. This is not an aesthetic statement, but one about the properties of the gas itself, in particular that we may regard it as being composed of particles that collide elastically. Departures from ideal behaviour are small, and it turns out that in large parts of the solar interior an ideal description suffices quite nicely.

Under ideal conditions temperature is proportional to the product of density and pressure. With this information in hand we can arrive at an inequality for the mean temperature in the interior. This is tantamount to performing a juggling act. The gas must be distributed throughout the interior in such a way that equilibrium is maintained. For different possible mean interior temperatures one is fixing the density needed to

---

[3] Estimates from the current best solar models give a value a few hundred thousand million times the sea-level pressure.

[4] We ignore the possible impact of rapid internal rotation, which can provide additional support from centrifugal force. The subject of rotation is considered at length in Chapters 9 and 10.

support the weight of the overlying material. But one does not have an infinite amount of gas to play with – only that making up the Sun. It is then a case of seeing whether, when one reaches the deep interior of the star, enough atoms remain available to pack in to give the required density. If not, the mean temperature must be altered and one tries again.

This is all taken care of by the mathematical equations. These indicate the mean interior temperature is expected to be of the order of several million degrees centigrade.

Sir Arthur Eddington described this basic approach in erudite language in *Stars and Atoms*. He went on:[5] 'The mathematician can go a step beyond this; instead of merely finding a lower limit, he can ascertain what must be nearly the true temperature distribution by taking into account the fact that the temperature must not be "patchy".' These steps were first taken in the latter half of the nineteenth century. We may attribute the first model of the interior properties of the Sun to the American physicist J. Homer Lane. His work, presented to the National Academy of Sciences in 1869, was recorded in the *American Journal of Science and Arts* the following year.

Lane was interested chiefly in arriving at values for the temperature and density at the surface of the Sun, rather than within its interior. To estimate the temperature he started with recorded measurements of the so-called 'solar constant', the radiant power impinging (per unit area) at the Earth. To make use of this information Lane needed to know something about energy emitted by a radiating body.

The Sun is a near-perfect radiator and absorber of energy – a 'black body'. The rate at which energy is emitted by such a body is proportional to the fourth power of its temperature. Alas, Lane did not have the benefit of this result, for it was only published in 1879 (by the Slovenian physicist Jožef Stefan). He therefore had to make extrapolations from other work by Dulong and Petit, and Hopkins, and got a surface

---

[5] Eddington, *Stars and Atoms*, 1927, pp. 13–14.

temperature of 30,000 °C – not bad, considering the actual number is about 6000 °C.

To obtain an estimate of the surface density, Lane knew he would have to solve equations for the interior structure of the whole Sun. The extra information he needed to turn inequality into equality, and a usable expression, came from an assumption about how heat was carried up from the interior to the surface (where it would then be radiated away). A definite relationship between total pressure and density, independent of temperature, followed from this. It was then possible to derive estimates of the run of parameters in the interior, rather than simply set lower limits on them.

Lane opted for convection as the energy-transport mechanism and therefore sought a description of the interior of the Sun commensurate with it being in a state of 'convective equilibrium'. The idea of convective equilibrium comes down to us from Sir William Thomson (Lord Kelvin). If a bubble of gas or fluid is 'ideally enclosed in a sheath impermeable to heat',[6] it can carry its energy from one place to another by its physical motion. This is how convection transports energy.

Convection is familiar to us in many circumstances – for example, in a pan of water on a stove. Hot bubbles of liquid displaced upward from the bottom of a pan are buoyant and rise because they are less dense than the adjacent, cooler liquid in their new surroundings. This displacement will establish an upward-moving current of warm bubbles of liquid in the pan; meanwhile cool liquid from above will sink, forming a return flow. In this manner a circulation current is established – warm, buoyant fluid rises while cool, denser fluid sinks. The net result of these over-turning convective motions is an upward transfer of heat.

When a bubble holds onto its energy in the selfish manner described by Lord Kelvin it is said to behave adiabatically. Let us suppose a bubble of gas in the Sun is prodded upwards from its natural, equilibrium

---

[6] Sir W. Thomson (Lord Kelvin), *Mathematical and Physical Papers*, 5, 1911, pp. 254–83.

position. This motion is assumed to take place sufficiently fast that energy cannot be exchanged with the surroundings – so the change happens adiabatically. At the same time we do not want the bubble to rise too quickly – we want the pressure in the bubble to balance that of its newfound surroundings at each level of the rise. What fate then awaits the bubble?

If the bubble is heavier than its newfound surroundings it will sink back down. But if it is lighter it will continue to rise and is buoyant; this implies that in the original position the bubble was not in a contented, natural state of equilibrium. The layer is therefore unstable to convection, and the bubble can continue to rise, carrying its energy with it.

Lane had assumed this was the case throughout the solar interior and that an adiabatic description was therefore applicable. Under such circumstances the pressure of a gas is proportional to the density raised to the power of some constant. This 'adiabatic constant' takes a value that depends upon the thermodynamic properties of the gas.

Now for a little more jargon – an adiabatic change is also a type of polytropic change. This is one for which, when we supply some heat to a physical system, the change in temperature goes in simple proportion to what is added. The equation relating pressure and density is an example of the general polytropic form. Lane combined the equations of hydrostatic equilibrium, mass continuity and the polytropic pressure–density relation into a single function. This could then be solved to give the interior properties. Lane assumed the ideal-gas law held and that the gas would be composed of single atoms because the component atoms of molecules would have been 'torn asunder' by fierce collisions at the high temperatures of the interior.

There are three values of the polytropic index for which the combined equation has analytic solutions, i.e. ones that can be written as mathematical expressions. Alas, the value chosen by Lane did not correspond to one of these, and this forced him to solve the problem by more cumbersome numerical means. By doing so he estimated that the average thermal velocity of particles at the centre of the Sun was over

500 kilometres per second, a value implying temperatures of the order of several million degrees centigrade.

Unlike Lane, Arthur Ritter was more interested in structure throughout the interiors of stars. He did a similar analysis, independent of Lane, and his first results appeared in print in 1878. In this he derived the combined equation Lane had sought, but in the form we know it today. Ironically it does not carry Ritter's name. Instead it is called the Lane–Emden equation. In the late 1880s, Lord Kelvin applied his ideas on convective equilibrium to gas spheres. And then the work of these three pioneers was drawn together, and added to, by the aforementioned Emden. His extensive studies were summarised in the monograph *Gaskugeln* (Gas spheres).

From these early models – in which convection was assumed to be the dominant transport mechanism – it was possible to demonstrate that exotic levels of temperature, density and pressure were to be expected in the interior. As Sir Arthur Eddington noted in the 1920s,[7] what seemed to be absurdly high temperatures were not 'meaningless' but had to be 'taken quite literally'.

Eddington left as part of his legacy major improvements to the models. These took into account the fresh realisation that the dominant mode of energy transport was by electromagnetic radiation – photons, or 'aether waves' to use the language of the time – not convection. This demanded a theory to describe how the radiation interacted and travelled out through the interior to the surface. Karl Schwarzschild had already made progress in developing a theory of 'radiative equilibrium', which could be applied to the solar atmosphere. Thanks to this, and other work, it became possible for Eddington to incorporate the ideas into his models and a full treatment was given in his landmark book *The Internal Constitution of the Stars*. The results from the models suggested temperatures at the core of the Sun in excess of ten million degrees centigrade.

---

[7] Eddington, *Stars and Atoms*, p. 14.

In the new treatment an equation was needed to describe the flow of energy by radiation. The ease with which radiation can pass is determined by a property called the opacity of the gas. When the opacity is large greater obstacles are placed in the way of the photons and the resistance to the flow of radiation is high. It is under these circumstances that convection can become the dominant transport mechanism. However, in Eddington's models it was assumed that radiation carried the load throughout.

Several processes can distract and remove photons from the outward flow. Atoms can absorb the photons. This can either excite atomic electrons in their bound orbits to new energy states or remove the electrons entirely from the hold of the atom by the process of ionisation. The photons can also interact with the free electrons and lose energy in the process.

The heavier elements are a particularly important source of opacity. This is because they have large numbers of electrons and bound energy levels. These present a correspondingly large number of possible avenues down which the photons might be lured to interact. Careful atomic physics calculations are required to determine the opacity under the range of conditions found in stellar interiors; all the significant interactions that might take place must be covered. Opacity calculations have, understandably, come on in leaps and bounds since Eddington's time. As we shall see later on helioseismology has in effect done atomic physics by pointing the way to important improvements in these computations.

The 1920s and 1930s marked the major breakthrough in understanding what powered the Sun and it then became possible to incorporate into the models equations to describe the generation of energy.

Stars form from the collapse of large, diffuse clouds of gas. As these contract, energy from the gravitational attraction of the constituent particles is released. Half of this goes into heating up the cloud, while the other half is radiated away into space. At the turn of the twentieth century this contraction was regarded as the means by which stars derived their energy.[8] However, the timescales involved didn't quite add up.

Lord Kelvin recognised that the limited supply of energy made available by this process would result in the Sun completing its contraction on a timescale of a few tens of millions of years. This suggested a far shorter lifetime than the estimates then set by geologists from studies of the decay of radioactive elements in rocks. In spite of this apparent dilemma, Eddington noted[9] that Kelvin 'assured the geologists and biologists that they must confine their outlines of terrestrial history' within the few tens of millions of years derived from the contraction theory. However, by the 1920s Eddington and others had realised an alternative, and copious, supply of energy was available to the Sun – the energy tied up in its constituent atoms. Robert Atkinson and Fritz Houtermans first proposed the manner of its release in 1929.

In the very hot conditions found in the solar core ionised atoms zip around at extremely high speeds. Atkinson and Houtermans realised that in this frenzied state of thermal motion particles could smash into each other with such force that nuclear barriers could be overcome, allowing new elements to be formed by fusion. An important by-product would be the liberation of energy.

The basic paradigm became the following. As a foetal star undergoes contraction, temperatures in its core regions may reach sufficiently high values to trigger fusion. The extra source of energy then made available provides the additional support needed to stave off further contraction. But this is not just any type of fusion – it is controlled thermonuclear fusion. The rate at which energy is generated and the rate at which it is emitted at the surface delicately balance one another. The latter quantity, the luminosity, is one of the fundamental global, and measurable, properties of the Sun that any model worth its salt must match.

---

[8] Lord Kelvin, and German physicist Hermann von Helmholtz, calculated independently the energy available to a star from its gravitational contraction.

[9] Eddington, *Stars and Atoms*, p. 94.

The first physicists to write down the nuclear reactions for the Sun were Hans Bethe, Carl-Friedrich von Weizsacker and George Gamow. The Sun derives its energy not from a single reaction, but from a linked sequence or chain. Initially it was thought the dominant reactions were those that form a closed chain (a cycle) in which the principal players are carbon and nitrogen nuclei. However, after further careful work it became apparent this was not the case. The temperature and density were most likely not high enough, and the dominant chain was proposed as being one that begins with the reaction of hydrogen nuclei (protons). In addition to the liberation of energy, the reactions form heavy elements from lighter ones, the basic output of the first proton–proton chain being the conversion of four hydrogen nuclei into a single helium nucleus. There are two other important proton–proton chains in the Sun, involving reactions of elements up to boron.[10]

While physicists began to get to grips with fusion in the Sun, the internal composition of the solar gases remained a debated issue until after the Second World War. In his memoirs[11] the eminent Cambridge astrophysicist Fred Hoyle recalls talking with Eddington in 1940. Both then still held to the belief that most of the Sun was iron (about sixty-five per cent by mass), the remainder being largely hydrogen. As Hoyle noted, 'nobody, to my knowledge, believed in the second solution' of a predominantly hydrogen and helium composition. The solar spectrum was certainly full of lines formed by iron. However, a few years later the hydrogen and helium models began to hold sway. Observations of Fraunhofer lines and analyses of these data had by then established that the atmospheric composition, at least, was dominated by hydrogen in particular, together with helium, and that the heavier elements were present in far sparser quantities.[12]

---

[10] These heavier elements are forged in extremely tiny quantities.

[11] F. Hoyle, *Home Is Where the Wind Blows: Chapters from a Cosmologist's Life*, (Mill Valley, CA: University Science Books, 1994), pp. 153–4.

[12] The important pioneers of this work were Cecilia Payne, Henry Norris Russell and Bengt Strömgren. Payne's thesis received plaudits as 'the most brilliant PhD thesis ever written in astronomy'. Russell lent his name to the 'Hertzsprung–Russell diagram'. On it,

The models of Eddington had assumed radiative equilibrium held throughout the entire volumes of stars. Convection did not feature. The question of whether or not parts of the solar interior might support convection was looked at in greater detail in the 1930s.

As we alluded to earlier, when the opacity is high, convection can take over, and this is what happens in the outer layers of the Sun. For a bubble of gas to be buoyant and rise, it needs to be lighter than its surroundings. We know the Sun is stratified under gravity – further out from the centre the gas is more rarefied. Provided the density gradient the bubble encounters, as it moves adiabatically, is steeper than that in the immediate surroundings it can maintain its buoyancy. A layer in which this is true is then unstable to, and will support, convection. We can also describe this instability condition in terms of a gradient of temperature – the gradient seen by the adiabatic bubble must now be shallower than that in the surroundings if it is to stay hotter and so be less dense than them.

Karl Schwarzschild introduced the convective instability condition into astrophysics.[13] The process that acts as the trigger to satisfy the condition in the outer layers of the solar interior is ionisation. The hungry atoms gobble up the radiation. Since fewer photons can pass unhindered through the layers in question the radiative temperature gradient steepens to compensate for the increased opacity. This tips the balance, and convection becomes the dominant mode of energy transport.

Ludwig Biermann was the first scientist to attempt to describe the transport of energy by convection in the layers near the solar surface. He, and others, soon found this was far from straightforward. He based his treatment on something called mixing length theory, which had been developed from observations of convection in the rather different physical conditions found in laboratory experiments. Convection is a

---

the brightness of stars is plotted against their temperature. This separates out stars of different types, revealing, in essence, a full life history. An example of such a diagram, for a cluster of old stars, is presented in Chapter 8 (Figure 8.2).

[13] Schwarzschild did so from the meteorological literature.

horrendously complicated process. The mixing length approach in essence says something like this: Fair enough, we cannot describe in detail what is happening on all the different scales; so instead we will seek solace in a description of the average behaviour. This zeros in on a characteristic length scale – the mixing length. Parcels of gas are assumed to travel this distance before they lose their excess heat and dissolve into the surroundings.

At first it was assumed the mixing length, and the depth of the convecting layer itself, matched the horizontal dimensions of the signature of convection observed on the surface of the Sun. This signature manifests as a honeycomb-like pattern of cells. The centres of the cells, or granules, are bright and contain hot, up-swelling material. The dark, inter-granular lanes contain cooler, downward-moving material that, having dumped its excess heat, is re-directed to form the return flow of the convective circulation. The typical cell dimensions on the surface are of the order of 1000 kilometres. It soon became apparent that the sub-surface convection zone could not be quite so thin – by the early 1950s a value closer to 100,000 kilometres was felt to be more likely.

In the mixing length picture, the length scale of the convection in the Sun is governed by the characteristic length on which the density varies (the density scale height).[14] In deeper-lying, denser layers the scale height takes a large value; it decreases in size in the more tenuous layers nearer the surface. In order to reflect this variation the free parameter the models play with to tune the convection is given by the ratio of the mixing length to the density scale height. A single value of this *mixing length parameter* is then used to describe convection in any given star. The bigger the mixing length the more efficient is the convection. Put simply, more energy can then be carried by the motion.

In layers many tens of thousands of kilometres beneath the surface of the Sun, the typical values of density and temperature encountered make

---

[14] The pressure, as opposed to the density, scale height is actually used to fix the mixing length in models. This is because it is more readily obtained. In most cases it makes little difference which is used.

the convection reasonably efficient. The temperature gradient need only then exceed the adiabatic value by the tiniest of fractional amounts to carry the required amount of heat. What is more, the description of the convection cares little about the size of the mixing length parameter. This is not the case in the layers immediately below the surface. Here, the lower density and temperature make the convection far from efficient. There is less material here to physically transport the energy. This forces the temperature gradient to become noticeably steeper than the adiabatic value. The resulting gradient is given the gaudy label 'super-adiabatic'.

In order to predict the actual value of this gradient, careful tuning of the mixing length parameter is required. Although the convection zone may extend over a hundred thousand kilometres or more, the choice of an appropriate value for the mixing length parameter is dominated by matching what happens in the tricky super-adiabatic layer, which occupies a mere 150 or so kilometres immediately below the surface.

This choice also has direct implications for the predicted depth of the convection zone. A larger mixing length parameter implies a deeper zone.

By the 1950s solar models had reached a new level of complexity. With equations now included to describe the generation of energy by nuclear fusion, evolutionary sequences of models could be computed to follow the Sun as it grew up. In the 1960s more refined treatments of convection were implemented using the mixing length approach. Advances in computer technology were an immense aid. The philosophy adopted in building and evolving these solar models was much the same then as it is now. As we shall discover, the major differences have come in changes to the various ingredients that go into each part of the process. Let us go over briefly what is involved.

An embryonic model Sun is evolved from the point at which it has just come of age as a star – that is, it is just beginning to fuse hydrogen into helium in its core. It is then said to be a 'zero-age' Main Sequence star. The time the star then spends expending the reservoir of hydrogen fuel in its core constitutes its Main Sequence lifetime. For a star like the

Sun this takes about 10,000 million years. A sequence of evolutionary models is computed that follows the Sun from the zero-age up to the present age of about 4600 million years. This current-age estimate is robust and comes from dating of meteorites. The mass of the Sun can be calculated from planetary orbits and is usually assumed to remain fixed throughout the Sun's lifetime.[15]

Model construction is an iterative process. Models are built shell by shell, onion layer by onion layer. The gravitational forces within any given shell have to balance the difference in pressure across the shell (the latter dominated by the contribution from the gas pressure). The balance of energy into and out of each shell has to be considered – this needs to take into account not only energy generated in the shell by nuclear fusion, but also any changes that arise as the internal structure slowly alters during evolution. The composition change that drives the evolution – the conversion of hydrogen fuel into helium in the core – is key. The equations to describe fusion need information on the rates at which the various nuclear reaction chains occur. The transport of energy has to be fully described. Throughout most of the interior the dominant mode is by radiation. Estimates of the radiative opacity are therefore needed. In the outer layers convection takes over, and mixing length theory is applied.

The goal of any evolutionary model is to match, at the current age of the Sun, two observed gross properties. These properties are the luminosity – the total energy radiated into space every second – and the size (radius) of the Sun. The modeller usually has two free parameters available to them and the aim is to tweak these so the luminosity and radius come out right.

We have already mentioned one of these – the tricky mixing length parameter. Remember – this determines, very sensitively, the depth of the convection zone. From one standpoint the mixing length parameter may

---

[15] As we shall see, some later treatments have taken into account the modest amounts of mass that may be lost, and others material that might be taken on board (accreted).

be regarded as serving the purpose of tuning the radius of the model star to the correct value. However, once fixed, this does of course control the temperature gradient in the model convection zone, which in turn fixes the efficiency of the convection. The need to serve two purposes at once – returning the observed radius while matching the real temperature gradient in the convection zone – suggests the possible pitfall may always be lurking of getting one right but not the other. If one is interested only in determining the depth of the convection zone then mixing length theory, or whatever else one might choose to apply, becomes irrelevant. The radius of the Sun is known by observation. So, the problem is reduced to one of finding the radius in the interior at which convection takes over. A detailed theory of energy transport by convection is not needed to do this. But – if one wishes to compute a full model of the whole star, one must of course confront the problem of modelling the convection.

The other free parameter controls the abundance of helium in the star. The fractional abundances by mass of hydrogen, helium and heavier elements must all add up to unity. Estimates of the abundances of the heavy elements (in units of the hydrogen abundance) can be made observationally. These come from scrutiny of the numerous Fraunhofer lines made by the elements, allied with some theoretical modelling to turn these spectroscopic observations into abundances. The estimates that result are only directly pertinent to the photospheric layers at the surface, the layers within which the lines are formed. The assumption therefore has to be made that the estimates also provide a reasonable proxy for the abundances deeper down. This may not be the case in some evolutionary scenarios (as we shall discover in Chapter 6).

If it were also possible to obtain a robust estimate for helium we would now have the abundances sewn up. Alas, matters here are not so simple. The solar helium abundance cannot be inferred directly. Very high temperatures are required for helium to produce atomic transition lines in the solar atmosphere. Conditions like this are present in the most tenuous parts of the outer atmosphere, or corona. However, the low-density conditions in the corona mean it is very hard to measure reliably the amount of helium producing the lines. Much lower down in the atmosphere, where the Fraunhofer absorption lines are formed, the density is higher and many of the complexities encountered in the corona are removed. In spite of this the temperature here is too low to excite helium. No lines are seen, and no measurement is possible.

Determination of the helium abundance has rich cosmological significance. The Big Bang produced hydrogen and helium (and small amounts of lithium). The heavier elements have been made by nuclear reactions in successive generations of stars.[16] Although fusion has also increased the amounts of helium, the overall change over time is predicted to be small in comparison with the original quantities forged in the Big Bang.

To obtain some kind of estimate of the helium abundance, and a rough guide for the Sun, astronomers could look further afield. Take, for example, a cluster of stars thought to have a similar age to the Sun. Clusters form from large clouds of gas that fragment as they collapse, so that all the stars in a cluster share a common origin. One assumes it is possible to find in the cluster stars that look like the Sun, as well as stars whose surface temperatures are hot enough to allow their helium abundances to be measured directly. Once obtained, an estimate from this second cohort is assumed to provide an abundance estimate for the Sun-like stars – recall that all members of the cluster formed out of the same material. Finally, one infers that the abundance in the atmosphere of a star that looks like the Sun is the *same* as that of the Sun. This approach suggested a possible helium abundance for the Sun of about twenty-five per cent by mass.

The helium abundance is taken as a free parameter in the models. It serves the role of tuning the luminosity to the correct value. An increase in the helium content raises the mean molecular weight in the interior. This increases the temperature needed to maintain the star in hydrostatic equilibrium, which in turn pushes up the luminosity.

[16] The seminal paper on this is E.M. Burbridge, G.R. Burbridge, W.A. Fowler and F. Hoyle, *Reviews of Modern Physics*, 29, 1957, p. 547; see also the short overview: F. Hoyle, W.A. Fowler, G.R. Burbridge and E.M. Burbridge, *Science*, 124, 1956, p. 611.

Evolutionary sequences, made in the 1960s, constituted the first example of what we now call standard solar models. Although the state of play of solar modelling was therefore fairly mature, there was a whole series of questions about what the Sun really looked like inside. What was the actual profile of temperature? Just how deep was the convection zone? What was the helium content? There was then the issue of whether or not extras needed to be added into the models – so-called non-standard processes.

There were also the things the standard models did not take into account. These models ignored the effects of rotation throughout the interior – this was assumed not to be rapid enough to affect significantly the balance of forces. But separate efforts to model the rotational, or dynamic, evolution of stars like the Sun were already ongoing. What did the pattern of rotation really look like in the interior?

The standard solar models also ignored the effects of magnetic fields. There remained the question of how these were generated, how they permeated the interior and how the resulting, dramatic, visible manifestations of this magnetic activity – spots, flares and the like – came about.

And, by the mid 1960s, there was an additional huge headache for the solar physics community to confront, which had potentially dramatic consequences for the standard models – the solar neutrino problem.

Neutrinos are sub-atomic particles. Their existence was postulated in order to satisfy conservation of energy and momentum in experiments involving a type of nuclear reaction called beta decay. This converts sub-atomic protons into neutrons. Neutrinos interact only very weakly with other matter, making them incredibly hard to detect, and were for many years taken to have no mass at all (like photons).

Beta decay plays an important role in the nuclear fusion reactions in stars, by which neutrinos are released in copious quantities. Although neutrinos can come in three so-called 'flavours' – electron, muon and tau – only the electron type is liberated by the reactions in the cores of stars. The neutrinos are also released with different energies, depending on which part of the fusion reaction chain they come from.

The possibility of being able to detect the solar neutrinos took an important step forward when, in 1958, it was discovered that the probability for one of the reactions in the solar chain occurring was about a thousand times higher than had previously been thought. Two solar physicists, Willy Fowler and Al Cameron, realised immediately the significance of this and ventured that it might then be feasible to detect neutrinos arising from reactions higher up the chain. Ray Davis and John Bahcall were the two scientists who worked to turn the idea into reality.

John Bahcall entered solar physics at the start of the 1960s. At the time he was making a study of the timescales on which beta decay occurred in the harsh conditions found in stellar interiors. His results, published in 1962, had important implications for the aforementioned seminal paper by Burbridge, Burbridge, Fowler and Hoyle. The latter had used decay rates from experiments performed in the benign conditions of a laboratory. Bahcall found that the rates were sometimes altered appreciably if the reactions took place under the radically different physical conditions inside stars. It happened that Willy Fowler was chosen to peer review Bahcall's article. As a result of reading the paper Willy did two 'very characteristic' things (as John put it). First, Willy offered John a position with his research group at the California Institute of Technology ('CalTech'). Given the tremendous solar and stellar expertise resident in the Kellogg Laboratory at the time this was a fantastic opportunity for John. Second, Willy sent a copy of the paper to Ray Davis.

Davis was giving serious thought to the practicalities of detecting solar neutrinos. In particular, he was pondering a part of the chain that made neutrinos from beryllium nuclei. His idea was to detect these neutrinos through a reaction they could have with chlorine atoms, which would convert the chlorine to argon. However, because of the weak interaction of the neutrinos with matter vast quantities of chlorine would be needed to give even a few potential interactions and countable argon atoms. Ray hit upon the idea of using chlorine in solution – perchloroethylene, or cleaning fluid to you and me – housed in a huge underground tank that would act as a solar neutrino detector. However, to take this idea further he needed a robust estimate of the number of neutrinos the solar reaction would produce.

After reading the forwarded paper, Ray realised John could do the required calculations. He therefore wrote to John and asked if he would do just that. John appreciated the rich insight an observation of this type would offer and so quickly accepted. He recalled then spending the next six months carefully working through the calculation.

A vital element of the scientific armoury John would need to bring to bear to make a robust estimate was a good model of conditions within the solar interior. Specifically, he would need reliable estimates of the run of temperature and density. Since it was known the reaction rates were extremely sensitive to the temperature, even the tiniest of discrepancies could have had dire consequences for the calculated value. It was therefore vital to construct as detailed and accurate a model as possible, and this became John's priority when he moved to CalTech.

John's task was helped by the fact that two astrophysicists resident in the group – Icko Iben and Dick Sears – had already been making stellar models. However, these were for studying stars in much later stages of evolution than the Sun. With the help of Willy, John was able to persuade them to adapt the codes so they could be used to generate data for the present-day Sun. Allied with the vital nuclear physics worked out by John, this allowed the first estimates of neutrino emission to be made.

The initial numbers proved to be something of a disappointment, for they suggested the quantity of neutrinos generated fell some way short of the level Ray felt would be required to make an experiment practical. The projected detection rate in Ray's proposed tank was too low by a factor of about ten. However, in the late summer of 1963 John realised he had missed something out of his calculations. He had failed to include several exotic ways in which the chlorine might be turned into argon. With these 'super-allowed' reaction transitions added in, the reaction rate leapt up by a factor of about twenty. The experiment was now on.

Ray published the first results from his huge solar neutrino observatory – located at the bottom of a shaft in the Homestake Gold Mine in Lead, South Dakota – in 1968. The first important finding was that he could indeed detect the solar neutrinos – this provided observational confirmation that fusion did indeed power the Sun. However, when the numbers of neutrinos were measured, Ray found their number fell short of John's prediction by about a factor of 2½.[17] When, after the accumulation of more data, it became apparent this was not an aberration of the experiment the community realised they had a problem on their hands. A solar neutrino problem – but what was the solution?

One possibility was that the temperature in the core of the solar models was too high. This would have given a prediction for the neutrino flux that was an overestimate of the measured value. Solar neutrino problem solved! But the temperatures could not simply be reduced arbitrarily. If this offered a way to a solution, there had to be a good underlying reason to alter the models. And so, over the years that followed, a variety of mechanisms were proposed that would drive down the core temperatures of the models and at the same time allow the models to retain the observed luminosity and radius. But the possibility remained throughout that the problem might reside elsewhere – in the physics of the neutrinos themselves.

Helioseismology was to provide the means to discriminate between the new and standard solar models, and pointed the way to the solution of the Solar neutrino problem. It has allowed us to probe the internal structure of the Sun – thanks to sound waves trapped in the star and the resulting resonant oscillations. It is to the basics of sound and resonance that we therefore turn next.

[17] Ray's observations and John's then up-to-date prediction both appeared in the same edition of *Physical Review Letters*, 20, 1968: R. Davis, Jr, D.S. Harmer and K.C. Hoffman, p. 1205; J.N. Bahcall, N.A. Bahcall and G. Shaviv, p. 1209.

3

HOW TO MAKE A STAR RING

A discussion of sound trapped in musical wind instruments serves as a useful analogy for introducing the fundamental principles of the music of the Sun. We begin with some basics about sound and about standing waves and resonance in simple pipes.

Sound is a type of wave, a disturbance that transports energy from one place to another. It needs a medium in which to travel because material must be moved to create the wave. Since sound travels by particles of matter interacting with one another it is described as a mechanical form of wave.

A sound wave is made when the particles of a medium, for example molecules in air or ionised atoms in the Sun, are displaced backwards and forwards. This creates regions with pressure elevated above or reduced below the ambient, or 'equilibrium', value to which the medium would settle in the absence of the wave. These disturbances are called compressions and rarefactions.

Sound waves are also longitudinal in nature. This can be illustrated with the help of a Slinky toy. Periodic, back-and-forth displacement of a coil at one end of the toy will cause a wave to ripple down the Slinky as each segment in turn pulls and pushes the next one along. This

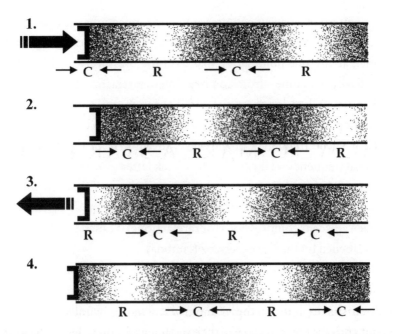

Figure 3.1. A piston (left-hand end of pipe) is used to make sound in a pipe from continual back-and-forth motion. In panel 1 the motion is caught at an instant when the piston moves into the pipe; panels 2, 3 and 4 follow in time sequence. Regions of elevated pressure (compressions) are labelled C; regions of reduced pressure (rarefactions) R. Arrows along the pipe indicate the direction of motion of the air molecules.

displacement, the mechanical motion creating the wave, is in the same direction as that along which energy is transported.

A piston at one end of a pipe in air can make sound (Figure 3.1). When the piston moves 'forward', into the pipe, it pushes air molecules in the same direction and creates a region of higher pressure. When the piston moves 'backwards', it creates a region of lower pressure (a rarefaction) and the air then rushes in to fill the void. If the piston continues to vibrate back and forth, the air molecules will be displaced, giving regions of elevated and reduced pressure. A wave is then observed to ripple out from the piston as molecules interact with their neighbours in the air. Not only does a pattern of adjacent compressions and rarefactions move down the pipe, provided the piston continues to drive back and forth in a periodic manner, at any one point along that path the pressure will also change periodically over time – from compression to rarefaction and back again.

The time taken for the wave to go through one complete cycle is called the period. Often, it is more convenient to talk in terms of the temporal or cyclic frequency of the wave, the inverse of the period. The temporal frequency – labelled by the Greek letter v (nu) – measures how many complete cycles of the wave there are per second, and has units of Hertz (cycles per second). The distance along the path of the wave between adjacent compressions (or rarefactions) is called its wavelength, and is labelled by the Greek letter λ (lambda).

For the moment we shall assume the waves we are dealing with are 'plane' in nature. This means they are confined to move in one direction only – for example down the long axis of a pipe with parallel edges – and do not converge or diverge. Furthermore, when plotted as a function of distance, or time, the pressure variations associated with these simple sound waves are sinusoidal. We could also consider the displacements of the air molecules. These are sinusoidal and occur at the same period as the pressure variations. However, the displacements are a quarter-wavelength out of step (out of phase) with the pressure changes. At an instant when the pressure is highest (or lowest) no motion or displacement is associated with the molecules. The displacement is largest when the pressure is at the neutral, ambient value when air is either rushing into a newly forming compression or out of a newly forming rarefaction.

The characteristic speed, c, at which sound waves move through a medium is governed by the ambient conditions. The speed, frequency and wavelength are linked by what is called a dispersion relation. The dispersion relation is simple and takes the form: sound speed equals frequency multiplied by wavelength, or in symbolic form: $c = v \times \lambda$.

The concept of a standing wave is the key ingredient of resonance in pipes. Take two sound waves that pass through one another in opposite directions. They are plane and have the same wavelength and pressure amplitude (by which we mean the difference in size between the highest pressure in a compression of the wave and the ambient pressure in the medium). As we will see very soon, reflecting just one wave from a boundary might produce such a combination.

At some point the compressions of one wave will coincide in space and time with those of the other. The waves are then in phase and the effect is to double the amplitude of each compression. The low-pressure parts of each wave also match and these regions in the air are further rarefied. Time passes. A quarter of a period later both waves have moved in opposite directions by a quarter of a wavelength. The compressions of one now coincide with the rarefactions of the other. The waves are now out of phase; the result is a net cancellation of any disturbance throughout the region occupied by the waves. Time winds forward a little more. After the passage of another quarter-period the waves are back in phase.

The overall effect is to give a strong sound wave that appears to stand still in the air. This is called a standing wave. It has the same period and wavelength as each of the travelling waves that combined, or 'interfered', to produce it. Regions in space occupied by compression will contain a rarefaction half a period later, and then a compression again a half period after that. Neighbouring regions that are in phase lie the distance of a wavelength apart. Air at the centre of one of the compressions or rarefactions never moves. These locations are called displacement 'nodes'. They are also pressure 'anti-nodes', because they lie where the pressure changes are largest. In between are found locations at which the pressure never changes but the displacement varies most strongly. These are pressure nodes and displacement anti-nodes.

A pipe resonates when a standing sound wave is set up within it. Here, one wave suffices to make the two needed to form the standing wave. This is achieved by reflecting the wave back on itself off the ends of the pipe. The wave is confined within a cavity. Attention must be paid to what reflection off each end of the pipe does to the wave. When the problem is couched in mathematical terms, the so-called 'boundary conditions' describe the effects of reflection. Depending on the physical nature of the end, we may, or may not, get a phase shift. When the end is a solid boundary an incoming compression is reflected as a compression, and the waves receive no shift. The end could also be open. This leads to partial reflection of a pressure wave, which once more allows a cavity to be formed – but now waves receive a shift of half a wavelength from what is termed 'free end' reflection.

The requirements needed to get a standing wave in a pipe, which must take into account the boundary conditions, are very exacting. For the majority of wavelengths the disturbances end up cancelling after multiple reflections. It is only at a specific set of wavelengths, which have a precise relationship with the length of the pipe, that the required reinforcement occurs. The standing waves produced at each of these discrete wavelengths are called resonant modes. When the pipe is fully open, air at the ends is essentially at the ambient (exterior) pressure. These locations therefore define pressure nodes and displacement anti-nodes (since the air is free to move). The first standing wave that will fit into the pipe – having the required pressure nodes at the ends – is one for which the wavelength is twice the length of the pipe. This is called the fundamental mode. It has the longest wavelength and period – the latter equal to twice the length of the pipe, divided by the sound speed – and therefore also the lowest frequency. All other modes must also have a pressure node (displacement anti-node) at each end of the pipe. The next mode has an extra node halfway along the pipe and is called the first overtone. The one after that is called the second overtone and has two extra nodes, and so on (Figure 3.2a). Up through this progression, to higher overtones, wavelengths get shorter, so periods get shorter and frequencies get higher.

When one of the ends of the pipe is closed, things are a little different. Air is now confined there, meaning a pressure anti-node (displacement node) is located at the closed end. The fundamental mode therefore fits a quarter-wavelength into a semi-closed pipe (upper part of Figure 3.2), and the fundamental period will be the same as that of a fully open

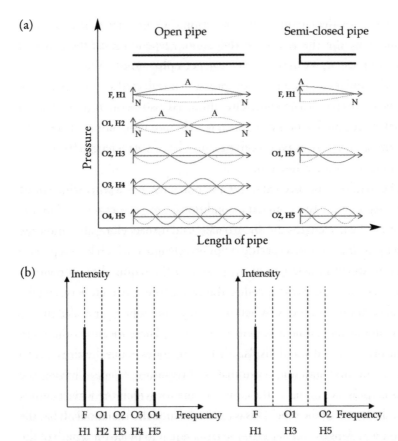

Figure 3.2. (a) Standing waves in a fully open (left-hand column) and semi-closed (right-hand column) pipe. The open pipe is twice the length of the semi-closed pipe. Panels show the variation of pressure along the length of the pipes for each standing mode. The solid line represents an instant when pressure anti-node locations are maximally compressed or rarefied. Half a period later, regions of compression are in rarefaction and vice versa. Pressure at this instant is shown by the dotted line. Modes are labelled F for fundamental, O for overtone (with number) and H for harmonic (with number). Even harmonics are missing from the semi-closed progression because only an *odd* number of quarter-wavelengths will satisfy the condition to give a standing wave (see text). (b) Conceptualised sound spectra for the open and semi-closed pipes (one twice the length of the other) from (a). Each mode is rendered as a dark spike in its frequency spectrum. The frequency scales are the same in both plots, the dashed lines serving as a comparative eye guide. The even harmonics are missing from the spectrum of the semi-closed pipe.

pipe of twice the length. The next mode up fits three-quarters of a wavelength, and the one after that five-quarters. An *odd* number of quarter-wavelengths must fit in to get a standing wave.

If we wanted to learn something about the resonant properties of a simple pipe, one option would be to record, and then plot over time, the intensity of the emitted sound.[1] However, it would be hard to tell which modes were contributing to the sound simply by listening to the output in this manner. Indeed, merely listening with the human ear can be misleading.

'Pitch' is the term used to describe how sound is perceived by the listener in terms of it being 'high', 'low', etc. For pure tones, given by a single wave or resonant mode, there is a clear correspondence between pitch and frequency – the higher the frequency, the higher the pitch.[2] However, matters are less trivial for the complex tones, produced by real pipes and instruments, containing the effects of more than one mode. One might think it would be possible to perceive the individual modes making up the combined output of an instrument. With some training this can be done, but the listener usually hears the pitch of the fundamental mode. Even when the fundamental is not actually sustained physically within the pipe, the ear generally perceives it as being present.[3]

To verify what is actually happening in the pipe it is more instructive to measure the *spectrum* of sound made by it. Here, the waveform characterising the pipe is split into its constituent components. The frequency spectrum is rendered by plotting frequency on the horizontal axis and the intensity of the sound at each frequency on the vertical axis. The waveform can be decomposed in this manner using clever mathematical techniques (like Fourier theory).

---

[1] Whether or not a mode sustained by a pipe-like instrument is heard depends on the transmission of sound out of its open end (or ends). When the wavelength of the standing wave is much longer than the cross-section of an end the wave is effectively confined. Flaring bell-shaped ends assist the transmission.

[2] The pitch of a pure tone also depends to some extent on the intensity of the sound.

[3] This is called the 'missing fundamental' effect.

For the moment we assume the interior of a pipe is awash with sound covering a range of many wavelengths. Waves with the correct characteristics cause it to resonate, and the fundamental and several overtones are established. The spectrum of the pipe is then dominated by these standing-wave modes. A plot will therefore consist of a series of strong peaks, each peak marking the location, in frequency, of an established mode. Information on the strength of the modes is given by the prominence of the peaks. Conceptualised examples for the open and semi-closed pipes introduced in Figure 3.2a are shown in (b) of the same figure.

The lowest-frequency peak is the fundamental, the next highest the first overtone and so on. In our one-dimensional, plane-wave example, these peaks are all spaced evenly in frequency – meaning the overtones are harmonics of the fundamental. In a fully open pipe, all harmonics are present. The fundamental is labelled the first harmonic, the first overtone the second harmonic, etc. (see the labelling in Figure 3.2a also). The frequency spacing is equal to the speed of sound divided by twice the length of the pipe. A semi-closed pipe half the length of a fully open pipe will have a fundamental mode at the same frequency. The spacing between consecutive overtones will be twice that of the longer, open pipe because the *even* harmonics are missing.

Pipes with cylindrical bores are a desirable form for a musical instrument because their modes are spaced evenly in frequency. Examples include a flute (fully open) and a clarinet (semi-closed). In reality, several factors contribute to small departures from this exact, harmonic pattern. Let us consider a few.

In our idealised description we took pressure nodes to be located exactly at the open ends of a pipe. The nodes are actually found just beyond the ends, and the size of this 'end correction' decreases with increasing frequency. The higher overtones in effect have a slightly smaller cavity.

The presence of a mouthpiece section (on, for example, a clarinet) or a bell at the other open end can affect the resonant frequencies of an instrument. Side holes also alter the shape of an instrument and the nature of the cavity, and are a key element of the wind instruments. The holes are not large enough to act like open ends. However, when exposed they alter the effective length of the pipe. By uncovering the side holes one by one – shortening the effective length of the resonant cavity – the frequency can be raised systematically.[4]

Then there is the matter of our having restricted the description to motion in one dimension only. In a real, cylindrical pipe, we must allow for motion in all three dimensions. The result is that there are also other families of modes that can be sustained. They come about from waves that execute circular motions in the transverse direction within the pipe, at right angles to the long axis. These extra modes, though only weakly excited in typical instruments, can be important from a musical point of view, and give a foretaste of the rich, three-dimensional spectrum of standing waves possible in a spherical body – like the Sun. We can make the jump from a cylindrical pipe to a sphere via another musically useful shape – the cone. Examples of conical instruments are the oboe and bassoon.[5]

A conical bore may be regarded as being a section from a sphere. A description of waves trapped within a cone requires that we treat the waves as being spherical in nature. As in a three-dimensional cylindrical pipe, we can have motion both along and at right angles to the long axis of the cone, meaning extra families of 'transverse' modes may also be sustained. When the resonant properties of a cone are sought, via the appropriate mathematical equations, the solutions yield two parts to describe each mode. Matters are more complicated than our simplified, sine-wave description of plane waves in a one-dimensional pipe.

---

[4] Six holes, placed suitably along a pipe, will suffice to allow the simple diatonic musical scale to be played. Say the fundamental of the fully closed pipe corresponds to middle C – a frequency of 262 Hertz. As each hole is exposed the progression will then take one up through middle D, E, F, G and A, finishing on middle B at 494 Hertz.

[5] The mouthpiece sections of these instruments alter the resonant properties compared with those of an ideal cone.

One part of the solution describes the mode properties in the transverse direction (parallel to the face of the cone or across the face of the sphere); this is the part not needed in the one-dimensional case. It is written in terms of 'spherical harmonic' functions. We shall show some pictorial examples later in the chapter.

Each mode given by the first part will have its own set of overtones, described by the second part. This second part carries information about what the standing waves look like along the long axis of the cone, and is therefore the part analogous to the solution for an idealised, one-dimensional pipe. However, whereas the one-dimensional solutions were simple sine waves, here they must instead be written in terms of Bessel functions.[6] The resonant frequencies given by these functions are not spaced exactly evenly, so the spectrum is not strictly harmonic in character. The frequencies can be described approximately by an analytic expression with an even spacing – this approximation gets better for the higher overtones because the spacings get progressively more uniform (so the approximation is 'asymptotic').

As in their cylindrical cousins, the transverse modes are not normally sustained to any appreciable level in typical conical instruments. However, they can become important in the conical-like geometry of a strongly flaring bell.

We might consider different types of conical bore having combinations of the following attributes: those with an open or closed large end; those complete to the tip; those with a truncated narrow end, which may or may not be sealed. These variations of structure lead to changes in the boundary conditions and differences in the resonant properties.

Let us zero in on a few specific examples, beginning with a perfect cone open at the large end. We can compare its frequencies with those of different types of one-dimensional, parallel pipe by ignoring the transverse modes. A perfect cone resonates at approximately the same

---

[6] A complete mathematical 'wave equation' for a cylindrical pipe is written in cylindrical polar co-ordinates, and its solution is also given in terms of Bessel functions.

frequencies as a fully open parallel pipe of the same length.[7] When no longer complete to the tip, but still open at both ends, the resulting 'frustrum' retains the same resonant properties (for the same slant length). However, when the narrow end is sealed – giving a truncated cone – things can be rather different.

If the part removed is a small fraction of the full length the frequencies are essentially unaffected. However, when the truncation becomes severe the resonant spectrum begins to resemble instead that of a semi-closed parallel tube. The smaller-wavelength, higher overtones will be more severely affected because they are less sensitive to the changing diameter of the pipe. The amount that must be removed to effect this change in behaviour – roughly one-quarter the length – depends on the overtone number of the mode.

When we generalise to the case of a uniform sphere, we have the same two-part solution as for a cone. However, with the restriction imposed by the narrow bore of a real instrument removed, the transverse modes are potentially of greater importance. This certainly turns out to be the case for the lead character of our book – a rich number of these additional types of mode are sustained in the Sun.

Our aim in the remainder of Chapter 3 is to start to tie in the ideas introduced above to the resonance and oscillations of stars. We shall cover the basic principles and introduce relevant concepts and terms. In subsequent chapters we shall then delve more deeply into the detail of the solar oscillations.

As indicated in Chapter 1, the Sun was by no means the first resonating star to be observed. By the 1960s several classes of pulsating stars were known. The classical Cepheid variables provide a nice starting point for linking the material together, in that the resonant oscillations they support are of the simplest variety possible.

The prototype Cepheid is δ Cephei (from which the name of the class derives). Careful observations of its single component of pulsation, of

---

[7] The slant length of the cone is the correct measure to take.

a period of just over 5 days, date back to the latter part of the eighteenth century. Before we address the physical manner of its oscillation there is an interesting aside, which relates directly to the discussion in the last chapter about the age problem and the contraction hypothesis.

By the early twentieth century astronomers were able to use over one hundred years of observational data on δ Cephei as a means to refute the idea stars derived their energy merely by contraction. Given this star's estimated mass and radius it was possible to estimate the rate of contraction needed to power it. This suggested the radius would have needed to change by about one-third of one per cent over the period in question. However, observations of the pulsations demonstrated the period had remained extremely stable, thereby ruling out the possibility that the radius could have changed by the amount required to power the star.

In 1880, Arthur Ritter had shown that the natural, fundamental period of oscillation of a gas sphere is inversely proportional to the square root of its mean density (a point we shall return to below). The form of oscillatory motion is purely radial, or vertical, meaning the sphere preserves its spherical shape as it expands and contracts periodically. This is colloquially referred to as a 'breathing' mode of oscillation. In 1914, Harlow Shapley suggested Cepheid variables might be undergoing radial pulsations of this type. If the mass of δ Cephei had remained unaltered, the constancy of its pulsation period implied a negligible change in density and therefore radius – provided the breathing mode hypothesis was correct (which it was). The conclusion was a clear one. Sir Arthur Eddington noted, 'At least during the Cepheid stage the stars are drawing on some source of energy other than that provided by contraction.'[8]

The physical nature of Cepheid pulsations is such that standing waves are set up like those in a simple pipe. But the pattern of nodes and anti-nodes has to be extended into three dimensions – the solutions for a conical wind instrument and full sphere are clearly of relevance. Let us

---

[8] Eddington, *Stars and Atoms*, p. 96.

start with the manner of the excitation of the pulsations. By this we mean the way the sound waves are generated in the interiors of such stars.

In our previous discussions of waves in pipes we assumed, implicitly, there was some constant supply of sound to feed the resonances. The issue of how sound is generated and dissipated (or damped) is an important one. We need a mechanism to make sound waves covering a range of wavelengths. Those waves with the correct characteristics will then be selected out to form standing waves. But the standing waves will survive only as long as the supply lasts. If the sound, once made, carries on for a long time – we then say it has a long lifetime – the demands on how often new sound must be made will be eased.

In woodwind instruments sound is generated in one of two ways: by blowing across an open edge, for example in a flute, which creates a jet of air and turbulent circulations called 'eddies'; or by vibrating a reed, as in a clarinet. Both methods generate sound which excites the instrument into resonance.

The nature of tones produced in real instruments is such that usually the fundamental and several overtones are excited. The precise details depend on the complicated nature of the excitation process. This process also provides the key to how woodwind instruments can typically cover several octaves in range. For example, when a flute player blows harder, the frequency of the sound generated (the so-called 'edge tone') increases. If the player can raise the frequency sufficiently it will match that of the first overtone of the flute pipe more closely than that of the fundamental. The first overtone is then driven more strongly and becomes the dominant mode, doubling the frequency and thereby raising it an octave. This 'overblowing' is much harder to accomplish on a clarinet; a small hole placed not far from the mouthpiece helps to put more energy into the first overtone, which for a semi-closed pipe is the third harmonic.

A single blow gives a short-lived toot from a woodwind or brass instrument. The sound dies very quickly, and the resonance is heavily damped out. Repeated blowing is therefore demanded of the player.

When simple waves are damped in this manner, there is an effect on the properties of the harmonic spectrum. Plotted against time, the sinusoidal pressure variations die away, decaying on a timescale corresponding to the lifetime of the sound wave. The peak of a resonant mode made by a decaying wave then has some width in the spectrum. No longer a mere spike, the width of the peak is a measure of the lifetime and will increase under the influence of heavier damping.

In a Cepheid the site of excitation lies in the outer parts of the interior, in a layer where helium is being ionised for the second time (that is, the atoms are being stripped of their second electron). These circumstances create a huge valve for the flow of radiation, which powers the pulsations as the valve opens and closes. Eddington first proposed this idea of treating the star like a huge 'heat engine', but it was only in the 1950s that the Russian astronomer Zhevakin identified the aforementioned zones of ionisation as the likely seat of the pulsation.

Consider first that part of the pulsation cycle where the layer has moved inward and is maximally compressed. The valve then closes off. In technical terms this means the opacity of the layer has increased. Radiation (the steam of our engine) that would otherwise have escaped is prevented from doing so. Heat and pressure build up beneath the layer, and the layer is pushed outward, overshooting the radius it would otherwise have reached. Now fully expanded, the valve opens because the opacity drops. The layer is now more transparent to radiation. Energy can escape freely via the radiation, and the reduction in pressure means the layer contracts, falling inwards below the radius that would have resulted had the valve remained closed. In this manner the layer acts like a piston, and waves of sound are generated inside the star. Over successive cycles they build in size. Because of this tendency to want to grow Eddington called the motion 'over-stable'. Damping processes eventually slow the growth.

The characteristics of the driving layer have to be just right for the process to work efficiently. This means only some stars are able to support pulsations of this type. Stars are often categorised by plotting them on a diagram with a measure of luminosity on the vertical axis and a measure of temperature (colour) on the horizontal axis.[9] Cepheid variables fall in a narrow, vertical 'classical instability strip' in this so-called Hertzsprung–Russell diagram.

The interior of a Cepheid forms a three-dimensional cavity within which sound waves generated by the heat-engine mechanism can interfere to form standing resonances. The star may be thought of as producing resonances rather like a conical pipe – gas is confined at the centre (the pointed end of the cone) and free at the surface (the open end).

The sound waves move along radial lines in three dimensions. These lines project out from the centre of the star (along the long axis of conical sections) and intersect the surface at right angles. We then say all the motion is vertical. Since the layer of excitation is in effect a buried spherical annulus, which expands and contracts, the fact that waves have spherical symmetry is perhaps not surprising. When these waves interfere the resulting patterns therefore also have spherical symmetry, and the nodes lie on spherical surfaces. As the standing waves move the gas backwards and forwards the surface of the star is seen to contract as one, then expand as one – the star oscillates. This simple, spherically symmetric pattern is a breathing, or radial, mode of oscillation. The other families of modes we referred to in our discussion of three-dimensional cylinders and cones – those arising from transverse motion – are not excited to sufficient levels to be observable in a Cepheid.

By analogy with a fully open pipe, whose resonant properties a perfect cone follows, we might naively expect the fundamental mode to possess a frequency approximately equal to the (harmonic) mean of the speed of sound in the interior, divided by twice the radius of the star, and the overtones to be harmonics of this fundamental. In practice there are departures from these simple predictions because of the spherical shape of the star, which leads to unevenly spaced nodes, the variation of conditions throughout the interior (the stratification we referred to in our

---

[9] Recall Henry Norris Russell, Chapter 2. See also Figure 8.2.

discussion of the Sun in Chapter 2), which affects the propagation of the sound waves, and the more complicated boundary conditions. Nevertheless, the sound travel time across the star does give a rough estimate of the fundamental period and is inversely proportional to the square root of the mean density in the star.[10]

Cepheid variables show pulsation in one or two modes. These may be some combination of the fundamental and lowest radial overtones. The periods associated with the pulsations are on the order of days or weeks. This is to be expected based on the mass and size of these objects. The pulsations are very amenable to observation because the compressions bring about large changes in temperature and therefore also luminosity. When plotted against time the brightness variations are not sinusoidal but slightly distorted in shape. The amplitudes of motion are very large indeed. They can result in the radius of the star changing by several per cent during each cycle of pulsation. The velocity associated with the motion is typically around a few tens of kilometres per second.

How does the description differ for pulsations of the Sun? The modes of oscillation that have been observed on the Sun also result from interference between sound waves. But the sound is made in a very different way. Here, it is the turbulence that is a feature of the super-adiabatic layer at the top of the convection zone that generates the acoustic noise. But whereas in a Cepheid the waves build to large amplitude, they are

[10] The proportionality to density can be demonstrated as follows. First, a suitable average of the sound speed across the star would allow for a determination of the travel time. Now, the speed of sound in a gas is given approximately by the square root of the ratio of pressure and density. There is also a correction, the adiabatic constant, which we have seen depends on the thermodynamic state of the gas. By use of the equation of hydrostatic equilibrium it is possible to obtain an expression for the ratio of the mean pressure and mean density, given in terms of the mass and radius of the star (and Newton's constant of gravitation). A description of the mean speed of sound in the interior can therefore be given in terms of these basic quantities, and in turn be used to formulate an expression for the travel time, from which the square-root dependence on density drops out. It is perhaps worth adding that strict scaling of the fundamental period, from one star to another, does not depend on mean density alone; nevertheless, the simple square-root dependence does allow for a reasonable estimate of the timescales involved.

intrinsically limited to small values in the Sun. The following rough-and-ready analogy may help.

Take a bell. At one extreme we have Cepheid-like pulsation in which the excitation of sound proceeds by giving the bell a series of hard thwacks with a hammer. These must be made at just the right phase every period, like pushing a child on a swing. At the other extreme we have the much more gentle, solar-like pulsation. For this case imagine the bell enveloped in a moderate dust storm. The bell gets repeated, tiny kicks from the impacting grains, and these make it sound. But at the same time some of the grains will strike at just the right moment to damp out the motion. So, although the process of the grains striking the bell drives the resonance, the same phenomenon also acts to damp it out.

On the Sun the little kicks come from the convection, and this acts to damp the sound waves too. The processes involved in excitation and damping of sound are very complicated; much effort has gone into developing both analytic and numerical models of them. The simple point to be made here is that the amplitudes are tiny compared with those in a Cepheid.

Once made, the sound can interfere to form resonant standing waves in cavities within the solar interior. Radial, breathing modes of oscillation are observed, but the surface displacements associated with them are now measured in terms of metres, rather than the many tens of thousands of kilometres possible for a Cepheid. But the radial, vertical modes are not the only ones that get excited. Propagation is not all in the radial direction; it is possible to set up transverse, or non-radial, modes. The motions then have a sideways component in addition to the stronger vertical, up-and-down motion. One example is a mode for which opposing hemispheres contract and expand out of phase with, or in opposition to, one another. As the motion becomes more complicated in the next modes the patterns become increasingly elaborate and much harder to describe in words, but the spherical harmonic functions we mentioned earlier do the job mathematically for us. (See some examples in Figure 3.3a.) The Sun supports well over a million of these modes. Each type has its own set of overtones, like a pipe or Cepheid.

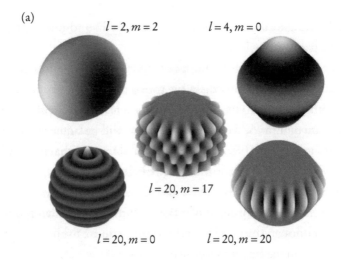

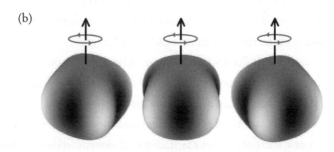

Figure 3.3. (a) Some spherical harmonic patterns, showing extreme displacements for each combination of *l* and *m*. The rotation (preferred) axis is vertical. The displacement has in each case been greatly exaggerated – with respect to the solar case – to reveal the detail of the patterns.

(b) Example of an *l* = 4, *m* = 4 mode at different phases of the cycle of pulsation. The graphical annotations above each sphere indicate the sense of rotation. In going, say, from left to right, the patterns are 'carried' around (advected) by this rotation.

Stars rotate. Although the rhythmic pattern of pulsation of the radial modes will look the same to any observer as the Sun (or a Cepheid) spins on its axis of rotation, this is not the case for their non-radial cousins. Here, the sound waves travel not only in and out of but also around the star.

If the motion of a wave has a component in the direction of the rotation, the rotation will help to carry the wave round the star. We say the wave is advected – the additional velocity component acts like a Doppler shift and the frequency of a mode formed by the wave will be higher than in the absence of the rotation (Figure 3.3b). But what if a wave, having the same basic properties, were to move instead in a direction against the rotation? It would of course have the frequency of its mode reduced. For the radial modes one simply cannot see this rotational advection. However, in the non-radial modes one can, and the size of the frequency shift will depend on the rotation in the layers the wave traverses.[11]

It turns out there are allowed orientations that lead to the formation of standing modes. Where in the absence of rotation there was a single mode, in its presence there will instead be several mode 'components', whose frequencies will depend on the orientation of their wave patterns with respect to the direction of the rotation, and of course the speed of rotation in the cavities within which they are trapped.

At the centre of the mode 'multiplet' will be a single component, whose frequency will correspond to that the mode would have had in the absence of the rotation. It is formed by waves whose motion is all at right angles to the rotation[12] – meaning it gets no shift.

The shifted components surround the central component in frequency and come in pairs (one shifted to a high frequency, one to a low one). Those formed by waves moving with the rotation are called prograde modes; those against are called retrograde modes. Components formed by waves moving parallel to or against the full rotation will have the largest frequency separation. If there are allowed orientations in between, there will be other pairs with intermediate-sized shifts.

---

[11] The effects of centrifugal force are small in the Sun. A complete description also includes a small contribution to the frequency shift from the Coriolis force (of which more in Chapter 10).

[12] Formally parallel to the axis of rotation.

Measurement of the resulting frequency splittings of the non-radial modes therefore allows the internal rotation to be probed.

The patterns of oscillations are described formally by the spherical harmonics (Figure 3.3a). The surface patterns from individual excitations in the more complicated non-radial modes take on the appearance of a chequerboard stretched into the shape of a sphere. Adjacent patches mark out regions that are moving either towards or away from the observer. Over a full cycle of pulsation the displacement of each changes from being a depressed dimple to a jutting-out pimple. However, the resulting distortion of the Sun's spherical figure is tiny. Each mode has its own unique set of spherical harmonic parameters. The parameter set used in the mathematical expressions to fix the form of the pattern consists of two integers – l and m. To allow a formal identification of the mode an additional integer, n, is needed.

The radial order, n, measures the number of nodes in the vertical direction, like the overtone number for modes in a pipe. The radial modes only have nodes that lie on spherical surfaces, nested in depth. The non-radial modes have additional nodes, which criss-cross each of these surfaces, marking boundaries between the oscillating patches. They intersect the Sun's surface; the number of lines crossing the surface[13] is called the angular degree, l. The radial modes therefore have an l of zero. When l is greater than zero, we know the modes are split by the rotation into several components. This means we need another integer, called the azimuthal degree, m, to tag each of these split components, with their distinct patterns of oscillation over the surface.

While the radial modes penetrate the whole interior of the Sun the same cannot be said of the multitude of non-radial pulsations. As we shall see later, the higher the angular degree, l, the smaller is the volume of the interior that is engaged in pulsation. These volumes are confined progressively closer to the surface, sections of which therefore resemble truncated cones.

---

[13] This is the number of complete circles, in longitude and latitude.

The modes formed by sound waves are often colloquially referred to as 'p modes'. The 'p' is for pressure – to indicate that changes in pressure, which produce the sound, are the main force that drives the oscillations. However, these are not the only type of interior waves the Sun can support. The other type is called a buoyancy wave.

A buoyancy wave will be produced when a parcel of gas bobs up and down around a stable, equilibrium position. Although the resulting waves can propagate in the interior of the Sun and set up their own pattern of standing waves, there are several crucial differences from sound waves and p modes. It is gravity that acts as the restoring force for the motion – hence the description 'gravity [g] waves'. As well as being displaced, gas parcels will in general (but not always) experience some compression.[14] However, this is less severe than in the acoustic case. A lot of the motion associated with the gravity waves is horizontal, not vertical. This is because parcels of gas have to push material out of the way as they oscillate up and down. As a result there are no radial internal gravity modes.

The characteristic periods of gravity waves are longer than those of the acoustic waves. What is more, unlike their acoustic cousins, gravity waves are rudely turned back when they attempt to enter the convection zone. In order to have periodic, wave-like behaviour, material has to be able to bob up and down around the stable location. But in the convection zone this oscillation is not possible. Material is instead press-ganged into circulating convection currents, which result from the inherent buoyancy, and rhythmic displacement is no longer possible. We shall discover this has had dire consequences for attempts to uncover the signature of deep-lying gravity modes.

Before we conclude the chapter a few comments on how the pulsations are modelled are in order.[15] To calculate the resonant properties of the

---

[14] The exceptions are surface waves, analogous to ripples on the surface of water. The so-called 'f modes', which result from surface waves, are discussed towards the end of Chapter 6.

[15] A substantial literature on stellar pulsations already existed before the global nature of the solar oscillations was uncovered, thanks to the likes of Eddington, Thomas Cowling, Art Cox, John Cox, Paul Ledoux, Chaim Pekeris, Théodore Walraven and

model of a star the equations used to build it must be modified to incorporate Newton's second law of motion – layers of the interior are no
longer static.

Several assumptions can usually be made to keep the resulting problem as-simple as possible. First, the timescale on which energy is
exchanged by radiation in the interiors of stars is usually much longer
than that on which sound travels. Technically, this means the motions
associated with the oscillations occur adiabatically. This holds throughout most of the solar interior, only breaking down near the surface. In
the adiabatic approximation one does not, therefore, have to worry
about the equation of energy transport in describing the oscillations.[16]

Second, the speed of the motion associated with the pulsations is
often a small fraction of the sound speed. The displacements can therefore be treated as tiny wobbles about the static, equilibrium model. Since
the perturbations are so small their description can be kept as simple as
possible.[17]

We are now in a position to think, in very general terms, about what
the sound spectrum made by a pipe, or a star like the Sun, might tell us.
The longer we make a pipe the lower will be the frequencies at which it
resonates. The frequencies, and spacings, are also affected by the speed of
sound in the pipe. This depends approximately on the square root of the
ratio of the temperature of the gas to its mean molecular weight, or
equivalently (from the ideal gas law) the square root of the ratio of the
gas pressure and density. An increase in the temperature of the gas will
raise the sound speed and the frequencies will increase in direct

Charles Whitney. For those with a strong theoretical inclination, the following is an
excellent text: J.P. Cox, *Theory of Stellar Pulsation*, (Princeton, NJ: Princeton University
Press, 1980). An indispensable reference for the Sun-like pulsations is J. Christensen-
Dalsgaard, *Lecture Notes on Stellar Oscillations*, 5th edn (Aarhus: Institut for Fysik og
Astronomi, Aarhus Universitet, 2003), <http://astro.phys.au.dk/~jcd/oscilnotes/>.

[16] Full non-adiabatic models are nevertheless sometimes computed, in particular for
investigations of the excitation and damping mechanisms.

[17] More formally, they can be modelled as linear perturbations, and higher-order
effects can, to good approximation, be ignored.

proportion to the change in speed. This is a problem that can bedevil an orchestra in a poorly air-conditioned theatre – as its players warm up, a steady rise in frequency will result for the wind instruments. A one-degree increase at room temperature will raise the frequency of a pipe by about one part in six hundred. Musical instruments are therefore made to hit a particular note at a given temperature.

If the air in a resonant pipe is replaced by another gas (or mixture of gases) the frequency will also alter if there is a change to the mean molecular weight of the medium. Were we to flush out air and replace it with helium gas – just over seven times lighter than air – the resonant frequencies would be raised by more than an octave.[18]

The frequencies of the modes in the solar spectrum are similarly affected by the internal conditions. The Sun's resonant spectrum exhibits near-regular patterns of overtones, like a pipe. Modes can have different cavity sizes in a woodwind instrument – for example, courtesy of the use of its side holes. In the Sun, the depth dependence of gas properties gives rise to a variation in cavity size from mode to mode. This means different modes probe different parts of the solar interior. At the same time, variations in properties close to the solar surface, strongly correlated with the 11-year activity cycle, give subtle and very small changes in the frequencies – like making small dints in the shape, or drilling tiny holes in the side, of a pipe. The peaks of the modes in the frequency spectrum also possess some width, on account of the damping. Whereas for wind instruments the characteristic time in which sound is damped in air is typically but a small fraction of a second, damping times for the waves in the Sun are typically of the order of days.

Nature has provided in the resonant modes of the Sun a wonderful tool with which to probe its interior. Many millions of modes of

---

[18] There is also a contribution from a change to the adiabatic constant, since nitrogen and oxygen molecules, which dominate air, are diatomic, whereas helium is monatomic. The humidity of air will also affect our orchestra in their poorly ventilated theatre. At room temperature the difference in frequency resulting from playing in fully saturated (one hundred per cent humidity) as opposed to dry (zero per cent humidity) air is about one part in three hundred.

oscillation are supported. Contrast this with a Cepheid, which may oscillate in only one mode – say, the fundamental. From the measurement of this frequency it is possible to estimate the mean density of the whole interior. Things get more interesting, and we get more information, if the star can sustain an additional overtone at an observable level. With two frequencies it becomes possible to make an estimate of the mass and radius of the star. But the information from two modes does not enable us to infer the entire interior physical profile. The Sun is a potential treasure trove in this regard. Each mode gives information on structure, and rotation, through the layers its constituent sound waves traverse – and there are lots of modes probing a vast gamut of interior ranges. From the spectrum of solar-like modes it then becomes possible to re-construct a profile, literally a map, of how properties like sound speed and density vary in the interior.

With that we complete our introductory chapters. With this background on the physics of the Sun and resonance, we are now ready to begin the story of how helioseismology unfolded.

4

THE STORY BEGINS

The period from 1960 to 1979 can now be regarded as the first phase of the history of helioseismology – its age of discovery, if you like. It took these twenty or so years to place the first, tantalising observations of the oscillations in their full and proper context. These truly were the first glimpses of a phenomenon that saw not just small parts of the Sun but the whole of its vast bulk gently oscillate.

There is a case for marking the beginning of the story – its preface at least – earlier in the mid 1940s. This is when scientists began to think seriously about the generation of sound near the Sun's surface. What happens to this sound is the key to our story.

Sound generation had been raised (notably by Ludwig Biermann and Martin Schwarzschild) as the origin of one possible solution to the 'temperature reversal' problem in the solar atmosphere. Observations had then only very recently established that temperatures in the outermost layers (the corona) greatly exceeded those found lower down at the visible surface (the photosphere). This raised the question of what could be powering this unexpected temperature rise. Biermann and Schwarzschild both hit upon the same idea, that of sound generated by turbulence in the outermost parts of the convection zone, just beneath the photosphere, as a possible source.

It is in the very outermost layers of the zone that conditions are most turbulent. The structure here is dominated by the honeycomb-like pattern of convective cells. The appearance of the hot cells, or granules, and cool lanes in between changes constantly. The surface appears to boil as the structure evolves, and cells are born and then die on timescales of the order of a few minutes. The key point here is that with turbulence come changes in pressure – and therefore sound. This happens over a spectrum of frequencies. Biermann and Schwarzschild suggested that energy taken up by these sound waves could be transported into the upper parts of the atmosphere by the waves themselves. The waves would therefore carry and dump mechanical energy to heat the corona. But this raised another question: was the solar atmosphere really awash with acoustic energy?

Our story begins proper in the mid 1950s, with a physicist from the California Institute of Technology named Bob Leighton. Though Leighton had received no formal training as an astronomer he nevertheless had a keen interest, which was soon to extend beyond more than being a hobby. It was around this time he began to develop an extremely novel technique for making measurements of magnetic fields on the surface of the Sun. Observations were made on the sixty-foot solar tower at the Mount Wilson Observatory. There Leighton made adaptations and improvements to the 'spectroheliograph' used by George Ellery Hale, some fifty years previously, to uncover the magnetic nature of sunspots.

Graduate student Bob Noyes joined Leighton in the autumn of 1958. Noyes had also received little exposure to astronomy and solar physics. His intention had been to pursue research in geophysics. But, having met Leighton, after teaching in one of Leighton's classes, Noyes began to help out with the magnetic field measurements.

Following publication of an extremely successful series of magnetic observations Leighton began to turn his thoughts towards looking at the ebbs and flows of the gas – the velocity field – in the solar atmosphere. He was particularly interested in the pattern of convective granules and the problem of measuring the lifetime over which individual cells survived before dissolving away. To collect the required data the spectrohelio-graph had to be modified slightly to measure the Doppler shift of chosen Fraunhofer lines, rather than the effects of magnetic fields on them.

The dark Fraunhofer lines that pepper the spectrum of sunlight can be identified with atoms of particular elements. These atoms must occupy a particular physical layer of the solar atmosphere at which the right temperatures and pressures are found for the line to be formed. Suppose the layer in question begins to move gently up and down. An observer who is marking the location in frequency of the line will see its position change in sympathy with the motion of the atoms. This is the Doppler effect, the same phenomenon whose classic textbook example is the change in pitch of a police car siren as the car speeds by an observer.

Leighton's set-up used a clever technique to record on photographic plates images whose intensities gave a measure of the Doppler shift.[1] It would scan a strip across the Sun – oriented approximately north–south – in one direction, and then re-trace its steps back in the other. It took a few minutes to complete an entire there-and-back scan. By comparing parts of each of these photographic Doppler plates corresponding to the same patch of Sun, it was possible to study how the velocity field varied over time. In the early summer of 1960 Leighton made the first series of scans, taking home the plates to develop in his own darkroom. What he found amazed him – and marked the beginnings of a new field of solar physics. When he came in the following day to announce his discovery, he told Bob Noyes, 'I've found your thesis!'

---

[1] The basic procedure called for the exposure of two plates – one for sunlight passed in a narrow band of wavelength on the blue side of the Fraunhofer line; the other for light on the red side. As the line shifted backwards and forwards, the light passed by the spectro-heliograph responded to the changing signal from each band. For example, when the line moved one way the signal would increase on one side and decrease on the other. As the line moved back the reverse would be true. By superimposing the blue and red plates – with one made as a negative – the Doppler shift contrast was enhanced. This 'differential' approach – taking and differencing signals from either side of a line – removed variations associated with intensity fluctuations that were common to both sides, leaving a plate dominated by the Doppler signal. This approach is common in instruments today that measure the solar oscillations.

Leighton had expected that for any location on the surface velocity signals separated in time by at least several minutes would no longer bear any relation to one another. This was the accepted timescale on which it was thought the granules lived. At separations in time of this, or more, newly formed granules, having an unrelated velocity pattern, would have replaced old granules. It was this signature Leighton thought would dominate the observed signal. But that is not what he found. Instead, when he looked at a given point on the Sun – and this worked right across the area scanned by the instrument – the patch moved one way for about 2½ minutes, and then moved back in the opposite direction 'to be what it was like before' after the passage of a full five (Figure 4.1). In short, small patches right across the surface were oscillating gently up and down with a period of about 5 minutes. Each of the 'velocity cells' was a few thousand kilometres in diameter – that is, each covered about two-millionths of the area of the solar disc.

At first Noyes did not believe it. But Leighton was adamant, insisting he had carefully checked the plates. Over the days that followed they were able to confirm the signal was genuine. Solar oscillations had been discovered.

Leighton made the first public announcement of the result at a conference in Varenna, in Italy, in August 1960. Further analysis of the 1960 data – and a subsequent set collected in 1961 – led to the publication of the discovery paper of Leighton, Noyes and George Simon in 1962.[2] Bob Noyes went on to make a detailed study of the oscillations the topic of his graduate thesis. He looked in detail at the differences given by observing the oscillations as velocity, or intensity, fluctuations; and also measured the period of the oscillations as a function of height in the atmosphere (by using different Fraunhofer lines).

[2] The paper – R.B. Leighton, R.W. Noyes and G.W. Simon, *Astrophysical Journal*, 135, 1962, p. 474 – was later judged to be one of the '53 most important papers to have appeared in the Astrophysical Journal and the Astronomical Journal in the 20th Century'. See *Astrophysical Journal American Astronomical Society Centennial Issue*, 525C, 1999, p. 962.

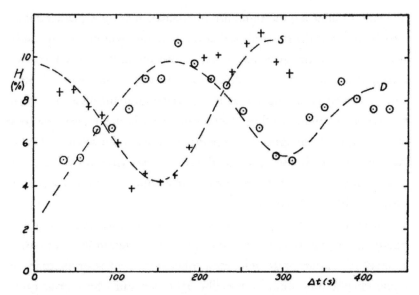

Figure 4.1. One of the figures from the Leighton, Noyes and Simon paper. It shows a measure of the Doppler shift (on the vertical axis) over time (about 400 seconds) for two types of exposed plate (see text), plotted as a function of time. The dashed lines indicate the implied oscillation. The data used to make this plot were collected on 10 June (curve described by circles) and 22 June (crosses) 1960. Figure from Leighton et al., op. cit., p. 474; reproduced with the kind permission of the American Astronomical Society.

George Simon was also a graduate student of Leighton. His field of study concerned the second big part of the discovery paper – the verification that a large-scale organisation of the granules, called 'super-granulation', which had first been observed by Hart in 1956, was not only present but also extended over the entire surface of the Sun. The discovery of oscillations soon received the support that is so crucial to any new observational finding: independent confirmation, by John Evans and Raymond Michard and by Bob Howard.

To the key players in the field it seemed likely there was a link between the small-scale 5-minute oscillations and the idea that sound waves could be present in the atmosphere. Soon, various mechanisms to explain the observations were being put forward that incorporated, in one way or another, the signature of acoustic waves. A further subset also entertained the possibility that the phenomenon might instead be associated with buoyancy waves in the atmosphere. This was the option favoured at first by Bob Noyes. It took a few years, and the right kind of data, to rule this out. The selection of acoustic mechanisms also began to be narrowed down. The three we now go on to discuss represent a good cross-section of the options the scientists were pondering in the 1960s.[3] Alas, none turned out to provide the correct description – but they do contain ingredients of the accepted explanation (which arrived at the beginning of the 1970s).

Two involved the trapping, by various means, of waves in the atmosphere. In one, Franz Kahn (in two papers submitted in 1961) discussed how the lower part of the solar atmosphere could act as a type of waveguide, shepherding sound waves like light waves in a fibre-optic cable. Here, the waveguide looked like two horizontal surfaces in the solar atmosphere. The idea was that waves trapped between the surfaces would skip repeatedly back and forth at shallow angles. Most of their motion would therefore be in a horizontal direction, parallel to the solar surface. Kahn then predicted that bundles of sound waves would be made to converge, as a result of repeated skips, at times separated by 4.8 minutes, very close to the 5-minute signal of Leighton and his colleagues.

To give the waveguide effect the mechanism relied on the presence of the temperature minimum at the top of the photosphere. This separates hotter regions below (in the interior) and above (in the chromosphere and corona). Waves are then effectively trapped within this minimum – or temperature well – by refraction.

Refraction is the mechanism that bends light when it passes through a prism, or gives rise to the mirage effect in a desert or above a road. To understand how a mirage is generated, consider a car travelling down a long, straight and flat road. A suitable combination of physical conditions

[3] A detailed perspective on the early stages of the field containing key references is: R.F. Stein and J.W. Leibacher, *Annual Reviews of Astronomy and Astrophysics*, 12, 1974, p. 407.

in the air above the road surface will make the phenomenon occur: there must be pronounced temperature gradient above the surface, air closer to the road being hotter than that further up.

Under normal circumstances one can see the road surface because it reflects sunlight (or the light from the lamps of a car). However, when the temperature profile is as described above rays of light heading downwards toward the road surface at a shallow, glancing angle never reach it. Instead, their path is altered so that they are bent upwards. This refraction takes place because the bottom of the wave front lies in the warmer air, closer to the road. Light travels faster in air that is hotter and therefore less dense. On the other hand, the top of the wave front is to be found at a greater distance above the surface, where the air is cooler and the wave speed correspondingly lower. Since the bottom of the front moves faster than the top, the wave bends around, or refracts. Were the resulting change in direction to carry the wave towards the observer in the oncoming vehicle, instead of seeing the road surface at the apparent point of reflection on the ground they would observe a mirage, a hazy image of the distant sky from where the wave actually originated.

It was the repeated refraction of sound waves either side of the temperature minimum in the solar atmosphere that acted in Kahn's model to trap waves. But the model drew criticism, largely on two counts. First, it was unclear whether it would actually give a sharp, well-defined signal across the photosphere. And, second, for the skip period to agree with observation the waves had to be predominantly horizontal. This was seemingly at odds with measurements of the 5-minute signal, which established the oscillations were much stronger in the vertical, radial direction, at right angles to the surface.

A second class of models took the atmospheric trapping idea further, to a point where standing waves were set up in a proper cavity, just like those set up in a pipe. Bahng and Schwarzschild proposed the formation of standing waves in the chromosphere, the layer immediately above the photosphere. Compressions and rarefactions of the gas in this chromospheric cavity would give rise to oscillations at its resonant period (or periods if several overtones could be maintained). If it could be shown this was close to 5 minutes the mechanism might be a serious candidate. But to form standing waves required a cavity. And a cavity needed a top and bottom.

We have already discussed a means to form the top for this cavity. The sharp increase in temperature one encounters moving up through the chromosphere, into the corona, acts very nicely as a reflector. Provided that waves are launched at an angle to the surface they can be refracted back towards the photosphere. An explanation of the proposed means of formation of the bottom of the cavity is a little trickier, and relies on an appreciation of characteristic speeds, times and length scales.

For a start, we know that for a sound wave to be a wave there must be a periodic pattern of compression; and the compressions needed to maintain such a wave travel at the local speed of sound in the gas. Let us train our attention on the temperature-reversing layer in the photosphere. Into this layer (from above) travels a very long-period wave. What happens when this period is much longer than the characteristic time it takes compressions to travel in this layer? The answer is: the wave is in trouble. The pressure changes needed to make this a wave cannot be maintained over a length of time matching its period. The compressions will be smoothed out because the gas can readjust on a time scale much shorter than the longer wave period. The disturbance can no longer travel as a periodic wave, down towards the photosphere, and is therefore reflected. The reflection forms the bottom of the cavity.

This is the state of affairs at the level where the temperature reverses in the atmosphere. There, the sharp decrease in the density gradient means the characteristic time for compressions to travel across the layer is very short. This timescale is most often expressed by helioseismologists as a frequency – the ominous-sounding acoustic cut-off frequency. The 'cut-off' part of the name tells us it is a *critical* frequency for acoustic waves because at frequencies below it the layer in question cannot support the waves. (Remember: a long period corresponds to a low frequency.) So, a local rise in the acoustic cut-off frequency can create a reflecting surface for sound waves. On the propagating side they can happily exist as compression waves. But does anything of a wave that is reflected in this manner get through to the other side of the surface? Surely, you say, the waves have been reflected so nothing can? This is not like a perfectly reflecting surface. However, even without any leaks something does actually get through. The gas on the other side responds to the compressions of waves on the propagating side, but the energy density dies off very quickly further away from the reflecting boundary. This dying signature is called an evanescent wave. Such a wave also pops up in the classical optics of light rays and prisms. We shall see that without these strange waves we would not be able to observe the 5-minute signal.

At the time Bahng and Schwarzschild wrote their paper, information on the physical conditions in the chromosphere was very sketchy. To the best of everyone's knowledge the cavity size suggested to meet the observed 5-minute period seemed pretty reasonable. With better knowledge it was later possible to show the cavity would instead have been expected to support modes with periods closer to 3 minutes. But the basic idea faced other objections. First, there appeared to be potential problems with waves staying in phase in the cavity because of large, and varying, non-uniformity in the high atmosphere. And, second, the model suggested a sudden fall-off in the size of the signal right at the bottom of the photosphere, just below the bottom of the cavity – not what had been observed.

The third of our mechanisms was called the granular piston. This idea relied on the impact of packets of hot gas at the top of the convection zone. These buoyant packets appear as the granules on the surface. When they reach the top of the zone, their upward motion does not simply grind to a sudden halt. Even though they are moving into a region that is stable against and so does not support convection, they still carry some excess buoyancy and overshoot into the photosphere above. These piston models – put forward by the likes of Jack Zirker, Friedrich Meyer and Hermann Schmidt – supposed the overshooting granules bumped and prodded the atmosphere from below, creating localised changes in pressure and therefore acoustic waves.

This mechanism, like the others, therefore took place in the solar atmosphere and did not extend into the interior below. It was not until the late 1960s that a scientist named Ed Frazier began to question whether this atmospheric interpretation was actually correct.

Frazier was working under the supervision of Louis Henyey at the University of California at Berkley. Henyey was an expert in building models of the solar interior and had a particular interest in the problems convection presented to the modeller. He shared these interests with another of his graduate students, Roger Ulrich (who will enter our story shortly). The mixing length approach to modelling convection was then, and still is, a far from satisfactory approach. Henyey felt that a way forward might be opened if one had better-quality, higher-resolution observations of the convection to study. While Roger Ulrich would pursue some theoretical aspects for his graduate thesis, Frazier was encouraged to make observations – and this he did, at the Kitt Peak Observatory. It was inevitable his data would also show the 5-minute signal and Frazier's findings would later inspire Roger to seek a theoretical explanation of the oscillations.

It was around this time the sophistication of the observations and analysis took a big leap forward. From the initial observations of the Doppler shift as a function of time, scientists had been able to make power spectra by Fourier-analysing the data.[4] The results then obtained showed the strength of the photospheric Doppler signal in different frequency bands. The next important step came in the mid to late 1960s: it involved turning these spectra from one- into two-dimensional plots. These new diagrams showed power in the signal as a function of frequency *and* wavenumber. Frazier was one of the first observers to adopt this approach to the analysis.

---

[4] The Leighton, Noyes and Simon paper had auto-correlations of the signal in time; Bob Noyes's thesis contained power spectra for observations made in Fraunhofer lines formed at different heights in the solar atmosphere.

The wavenumber of a wave is the inverse of its wavelength, times a constant (2π). It is also sometimes called spatial frequency because it has units of per metre. This is analogous to the relationship between period (measured in seconds) and temporal or cyclic frequency (inverse of period). Scientists like to work with wavenumber – designated by the letter k – as opposed to wavelength because plots of data are then rendered more clearly. Things tend to get scrunched up in plots against wavelength, but against wavenumber a large dynamic range can be covered more evenly. Before the 1980s, solar physicists also preferred to use angular, rather than temporal, frequency. The angular frequency of a wave – designated by the Greek letter omega, ω – is given by 2π times the temporal frequency, v.

By plotting power as a function of wavenumber as well as frequency – with ω on the vertical axis, k on the horizontal axis and height contours on the plot showing power – scientists like Ed Frazier could get a better handle on the physics they were observing. Wavenumber gave a measure of the horizontal length scale of any signal, in a direction across the solar surface. Important clues followed from how power in the resulting k–ω diagram was distributed. Phenomena involving wave-like motion would have been expected to arrange themselves into an organised pattern when plotted against the two parameters.

What did Frazier see when he did this? His observations revealed the 5-minute signal as two closely separated patches, or humps, of power in the k–ω diagram. For starters, this gave an estimate of the dynamic range of the phenomenon. Significant power was actually spread over a range of periods, from just under 4 up to about 7 minutes. This covered temporal frequencies from roughly 2½ to 4 milli-Hertz. Characteristic wavelengths ranged from roughly 3000 up to 10,000 kilometres.

The humps also occupied the upper left-hand part of the diagram, which meant they were consistent with being a signature of acoustic waves. Explanations relying on buoyancy waves could therefore be ruled out. This is because the k–ω diagram is split into an upper left-hand acoustic part and a lower right-hand buoyancy part. A special line separates the two parts. The slope of this line is the sound speed, *c*, for waves that travel horizontally in the layer in which the observations are being made. The waves obey the equation $\omega = c \times k$. This is the same dispersion relation we met in the last chapter – $c = v \times \lambda$ – now written in quantities relevant to our story here. For wavelength, λ, we have substituted wavenumber, k; and for temporal frequency, v, we have substituted angular frequency, ω. Each horizontal wavenumber has an associated frequency that obeys the relation, called the Lamb frequency.

Frazier's data had shown some mechanism was working to drive motions of the solar surface in a preferred range of frequency *and* wavenumber. Observations of a gas with completely random noise would have left a very messy, incoherent signature across the diagram. Furthermore, had the motions been dominated by a single-period component, power would have been concentrated on the single values of wavenumber and frequency characterising this motion. In spite of limitations in resolution, Frazier's observations would then have revealed a small spot, not the more extended patches of power that were seen.

To Frazier the fact the signal was strong at a handful of depths in the photosphere, right down to the bottom, seemed a crucial point. This was not what one would have expected from the then in-vogue 'granular piston', for which Frazier added, 'many oscillations were seen to develop without the aid of a convective perturbation'.[5] The conclusion he drew was that the signal originated, and penetrated, beneath the Sun's surface into the convection zone. Two scientists were soon able to offer, independently, a theoretical explanation that backed this up – the aforementioned Roger Ulrich, and John Leibacher.

Another important clue was already present in Ed Frazier's data, which both Roger and John latched onto. This was the double hump of power. Because they were arranged on top of one another in frequency

---

[5] E.N. Frazier, *Zeitschrift für Astrophysik*, 68, 1968, p. 345.

it was tempting to think the humps might be signatures of two overtones of a resonant cavity.

At this time John was a graduate student, working towards his PhD under the supervision of Bob Stein. He had been doing a series of one-dimensional calculations in an effort to understand the 5-minute signal. At first, he sought to understand Frazier's data in terms of the humps being standing waves trapped in an atmospheric cavity. A short paper, published in early 1969, elucidated a simple model of this. In order to make further headway John realised the model needed to be developed to make it more realistic. Meanwhile, prompted by the findings and conclusions of Frazier, Roger had begun to think about moving the cavity down into the interior. He soon found it was possible to set one up.

Sound waves launched into the interior at an angle to the surface would be bent around slowly by refraction on account of the increasing temperature. This refraction could form the bottom of an interior cavity. To form the upper boundary Roger could rely on the sharp decrease in density at the very top of the convection zone and base of the photosphere. The acoustic cut-off frequency would be very high in this layer. Waves coming up from below, with frequencies lower than the cut-off, would be reflected and their periodic disturbances injected back into the interior. The pattern of refraction and reflection would then repeat. Waves would be trapped in the interior and, with the right characteristics, set up standing wave patterns by interference with other suitably matched waves. The compressions and rarefactions of these standing waves would gently perturb the surface, and the signature would manifest as the 5-minute oscillation. This would be observable in the photosphere, above the bounded cavities, thanks to the evanescent waves.

John also reached similar conclusions. He recalls working feverishly in the basement at the National Centre for Atmospheric Research, in Boulder, on his new model, still looking for atmospheric modes. However, when he fired pulses of sound up from the interior he found resonant behaviour, and 'beautiful' 5-minute modes of oscillation, before the waves got anywhere near the atmosphere. Perplexed, John initially thought his computer code contained numerical gremlins. But, whatever he tried, he simply could not rid himself of this 'deeper-seated' 5-minute signal.

The crucial point in his uncovering these new resonances was that he had also avoided treating the bottom of the photosphere as a solid boundary below which waves could not travel. Although John initially imposed an arbitrary lower boundary, to seal the newfound interior cavity, he too had discovered that the observed 5-minute oscillations were entirely consistent with being the signatures of standing waves in the convection zone.

Both Roger and John found that an important consequence of the interior-cavity explanation was an expected pattern of characteristic ridges in the k–ω diagram, built from signatures of different families of overtones. To explain how these arise we can call on the waves-in-pipes ideas introduced in Chapter 3.

Waves travel in three dimensions within the solar cavity. They follow curved paths in the interior between bounces at the surface (Figure 4.2). To allow for this we need to talk in terms of a characteristic wavelength in both the vertical (radial) and horizontal (across the surface) directions. The vertical direction is analogous to the long axis of a pipe (the only direction that gets considered in a simple one-dimensional pipe), and the horizontal coordinate is like the transverse direction in a three-dimensional pipe.

The cavities Roger and John considered were shallow compared with the radius of the Sun, and the models did not take into account curvature arising from the spherical geometry. So from our stockpile of pipe analogies we might be tempted to think in terms of a pipe closed at the bottom (where waves are refracted) and open at the top (a 'free surface', where the steep density gradient gives rise to reflection[6]). The interference condition would then demand that an odd integer number of quarter vertical wavelengths fit between top and bottom.

---

[6] If one were to consider the spherical geometry, the shallowness of the cavities would make a severely truncated conical pipe the appropriate choice. This of course looks like a semi-closed pipe.

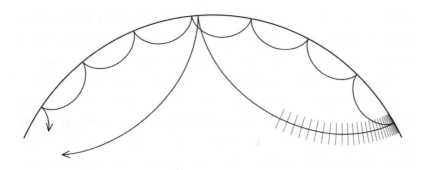

Figure 4.2. Paths followed by sound waves in the solar interior. The large arc defines the upper reflecting point for the waves – which coincides closely with the visible surface of the Sun – and the interior skips of waves are indicated by curved ray paths. Those for two waves are shown. Wave fronts showing locations at which the wave has the same phase are indicated for the trapped wave with the longer horizontal wavelength (short lines perpendicular to the ray path). The rendering here shows the curvature of the Sun, an element not included in the early models of the interior standing waves. Image reproduced with the kind permission of J. Christensen-Dalsgaard.

The full analogy does not quite hold. We know that in a simple pipe waves only get a phase shift when they reflect off an open end (a shift of half a wavelength). In the Sun waves get a phase shift at both the top and bottom of the cavity, and the boundary conditions are therefore different. It turns out that at the lower boundary waves receive a phase shift of a quarter of a wavelength. At the top the shift is roughly of the order of a full wavelength but varies with the frequency of the mode.[7] Changes in physical properties in the solar interior mean waves get phase shifted within the cavity too (contrary to what happens in a simple tube where the ambient conditions are assumed to be the same everywhere).

---

[7] Provided the horizontal wavenumber is not too high, all waves look similar near the surface, since they propagate very nearly vertically; this means, to a first approximation, the phase shift depends only on frequency.

Although this more complicated solar picture alters what is needed for the constructive interference condition to be satisfied, the basic idea of fitting waves into a cavity holds as true for the Sun as it does for a pipe.

Examples of paths followed by sound waves in the interior are shown in Figure 4.2. The more steeply a wave is reflected at the surface, the deeper it will penetrate the interior. Waves penetrate to a depth at which their horizontal velocity equals the local sound speed.[8] Waves that travel more nearly vertically are also those that have a longer horizontal wavelength (smaller horizontal wavenumber); their resulting internal trajectories mean they also travel a greater distance before reappearing at the surface than their smaller-wavelength counterparts.[9] So, cavity sizes vary depending on the characteristics of the waves. The larger the horizontal wavelength, the deeper is the cavity. This is in stark contrast to the simple pipe (side holes and end corrections aside) where the cavity is roughly the same size for all modes.

Now we come to the crux of what forms the ridges. At each horizontal wavenumber there is a family of solar modes that have different frequencies. They are the overtones of this horizontal wavenumber. The frequency of each overtone is determined by the interference condition in the vertical direction (the fitting of radial wavelengths from top to bottom in the cavity). As mentioned in the preceding chapter, the overtones are not quite spaced evenly in frequency (not harmonic).

The family of overtones of each horizontal wavenumber therefore occupies a vertical line of discrete points in the solar $k-\omega$ diagram. As one goes higher in frequency, one progresses to successively higher overtones. The next wavenumber along has its family and gives its own vertical pattern of points. Neighbouring patterns of vertical points accumulate to form curved, largely horizontal ridges lying above one another (see Roger's predicted ridges in Figure 4.3). Each

---

[8] We shall return to this point at the beginning of Chapter 8.

[9] The separation between consecutive reflections on the surface is not the same as the horizontal wavelength. But a greater separation goes hand in hand with a larger horizontal wavelength.

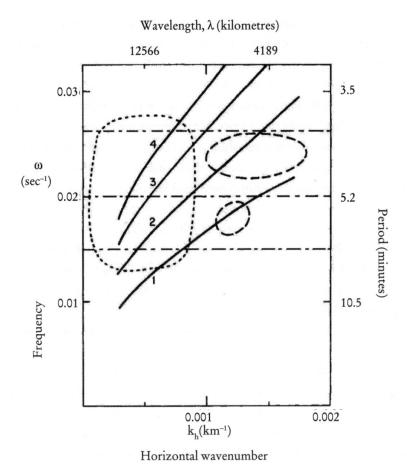

Figure 4.3. The *k*–*ω* diagram from Roger Ulrich's 1970 paper. The numbered lines mark the location of the first four 'ridges' as predicted by Roger's calculations. He also showed regions where oscillatory power had been seen by several observers. The large region bounded by the short dashed lines marks out power observed by Andrew Tanenbaum, John Wilcox, Ed Frazier and Bob Howard; the horizontal dot-dashed lines indicate observations by Gerard Gonczi and François Roddier (of whom more in Chapter 5); and the two ellipse-like features with long dashed lines are the humps of Ed Frazier. Adapted from figure in Ulrich, *Astrophysical Journal*, 162, 1970, p. 993; reproduced with the kind permission of the American Astronomical Society.

ridge is made up of modes with the same overtone number but a different horizontal wavenumber. This telltale ridge structure was to be the key to verifying observationally the theory of Roger and John.

John considered trapping in the outer 2000 kilometres or so of the zone. Roger used Frazier's observations to determine the range in horizontal wavenumber he would test. He ended up looking at modes with horizontal wavelengths up to about 20,000 kilometres; he found these penetrated to a similar depth beneath the surface. For the moment the models, and the overall picture, remained local in nature.

Roger first presented his ideas at a meeting in February 1969, and his theory was published in December 1970; John's work was published, with Bob Stein as co-author, within a few weeks of Roger's paper.[10]

In his paper Roger estimated the positions of the first four ridges (Figure 4.3). With his predictions to hand it was apparent that Frazier's humps of power possibly lay on the first and second ridges. But data of much higher resolution, in frequency and wavenumber, were needed to test the theory properly. Following the submission of his paper one of Roger's colleagues suggested he should add to it a recipe to give observers a clear idea of the data requirements needed to properly test the theory. The recipe, added in proof, called for two-dimensional $k-\omega$ data like those of Frazier's, but with superior resolution in frequency and horizontal wavenumber to separate out convincingly the anticipated ridges of mode power. Roger recalls that following the publication of his paper he had expected there would be a rush of observers attempting to do just that. However, the required observations failed to appear in the short term. This reflected the difficulty of the task in hand. Roger soon made the decision to begin his own programme of observations. At about the same time a German scientist in Freiburg, named Franz

[10] R.K. Ulrich, *Astrophysical Journal*, 162, 1970, p. 993; J.W. Leibacher and R.F. Stein, *Astrophysical Letters*, 7, 1971, p. 191. John and Bob's previous paper on the 5-minute oscillations (R.F. Stein and J.W. Leibacher, *Astrophysical Letters*, 3, 1969, p. 95) gives a flavour of how their ideas were developing. Roger's paper featured in the *Astrophysical Journal American Astronomical Society Centennial Issue*, 525C, 1999, p. 1199.

Deubner, set off on a similar course. We shall meet him again at the end of the chapter. By 1972 John had also initiated an attempt to find the ridges in collaboration with Jack Harvey and Oran White. Jack was on board because of his considerable instrumental skills, skills destined to give him a pivotal observational role in our story.

Jack got his first exposure to the field of solar oscillations in the mid 1960s. He was then a graduate student in Boulder and had been taking short-term observing jobs at the Mount Wilson Observatory in California. In the summer of 1964 he took a position to work with Bob Howard, one of the early observers of the 5-minute oscillations. Jack spent an enjoyable summer taking data, which were later analysed and published by Howard (a task Jack received full credit for in the resulting paper). However, Jack's proper research area was the study of magnetic fields on the Sun's surface. It would be a few more years before his research path overlapped once more with the solar oscillations.

With his doctorate thesis completed, and having now moved on to a position at the Kitt Peak Observatory, Jack became interested in the short-period oscillations (SPOs) reported by Bob Howard in 1967. Howard's data appeared to contain evidence for solar oscillations on a timescale of a few seconds – about one hundred times more rapid than the by then well-established 5-minute signal. Although Howard gave several reasons to doubt the solar origin of the SPOs, he could not find the hard evidence needed to back this up. Jack therefore set about this task and found the necessary inspiration in what seems the most unlikely of circumstances – a concert by the avant-garde classical composer John Cage.

As he sat listening to the organised cacophony, amused by the various sounds being used, something clicked – a picture and plan formed in Jack's head of how he could test the SPOs. The gizmo Jack built to do the job measured not only the Doppler velocity pattern but also the rate of change of velocity with position on the solar surface. The ability to measure this gradient was something new and, combined with simultaneous observations made from Mount Wilson and Kitt Peak, was the key to demonstrating the SPOs were generated by random noise in the Earth's atmosphere. Even though he now had the basis of an instrument ideally suited to studying the 5-minute oscillations, Jack's attention remained focused elsewhere. But this soon changed, thanks to his bumping into John Leibacher for the first time.

Jack met John at a meeting of solar physicists, held in Maryland, in March 1972. The topic of conversation soon turned to the 5-minute oscillations. John chatted over the ideas he had regarding their origin as standing waves in the solar interior. It was at this point Jack was able to offer his instrumental skills and with that they resolved to set up a programme to verify the theory observationally. The new collaboration, which included Oran White, was given the unforgettable acronym 'HOWL' by John – the first and fourth letters for his and Jack's surnames and the second and third letters for Oran's initials.

Jack spent the next few months setting up the necessary equipment and made two observational runs, one in September 1972, the other in October 1973. Unfortunately, the pressure of other work commitments meant the team did not get round to analysing the data. Although in some respects this was a missed opportunity, Jack now admits it is unlikely the data were good enough to resolve the sought-for power ridges. Nevertheless, the work undertaken for HOWL had laid the foundations for the huge observational contribution Jack would make from the late 1970s onwards.

Roger Ulrich's observational programme was well off the ground by 1973. His first task had been to acquire someone to spearhead the effort. The person he chose was Ed Rhodes. Ed first heard the words '5-minute oscillation of the Sun' while attending an astronomy class at the University of California at Los Angeles (UCLA) in the spring of 1970. The lecturer on this course was none other than Roger. Ed's recollection is that Roger worked through his theory of the trapping of waves in the solar interior in some depth and did not hold back on the mathematical content.

At the time Ed had just been accepted into the UCLA graduate school, his intention being to pursue a project in extragalactic astronomy for his PhD. As things turned out he was not able to do this, and so, faced with having to find a new topic, Ed soon found himself sitting in Roger's living room being asked whether he might be interested in verifying observationally Roger's theoretical ideas. Intrigued, Ed was on board for a project that would take several years of extremely hard work to bear fruit.

Roger and Ed's first objective was to find a suitable observatory and instrument to collect the data they needed. They first approached the San Fernando Observatory, and Ed took a part-time job there (formally with the Aerospace Corporation). Ed's principal role was to make magnetic field measurements of the Sun in support of the solar instrumentation on board NASA's manned *Skylab* space station. However, his position meant he could also work on the development of an instrument for the 5-minute oscillation project. After several months of work it soon became clear the instrument would not be up to the task (one for which it had not been designed). This left the question of what to do next.

The answer came partly by chance. Ed remembers attending a lecture given by Jack Harvey, who spoke at length about the programme of solar observations then ongoing at the Kitt Peak and Sacramento Peak observatories. Mention of one instrument in particular, set up in the large Vacuum Telescope at Kitt Peak, piqued the interest of Ed and Roger. When they approached Bill Livingston, who was in charge of the instrument, Livingston indicated it was probably not best suited to their needs. But – he knew of an instrument that was.

George Simon, at Sacramento Peak, was in charge of the Diode Array Magnetograph, the wonderfully named piece of kit Livingston had in mind. Simon seemed the ideal chap to approach. He was, after all, one of the authors of the 5-minute discovery paper. When approached by Roger and Ed, Simon was only too pleased to help. And so it was that Ed was able to set to work tuning the instrument.

The weather complied to allow good-quality data to be collected in stretches over a period spanning late December 1974 to mid January 1975. The instrument looked at a square patch covering less than one per cent of the surface area of the Sun. (The length of each side was about a seventh of a solar diameter.) Ed was then confronted with the huge task of turning the raw data into the requisite k–ω diagram. Since they were entering uncharted research territory, he did not have the luxury of being able to use already written procedures and computer code. Much of the analysis software therefore had to be written from scratch. As he continued to plough through the analysis, Ed also had to contend with preparations for the oral examination of his PhD, which took place in the summer of 1975. One of the members of his examining board was George Simon.

What Simon brought with him proved something of a shock and disappointment to Ed – a pre-print of a paper by Franz Deubner showing the sought-for Holy Grail: ridges! He and Roger had been beaten to the punch, but only just.

Deubner had been making high-quality observations of the 5-minute oscillations for many years, publishing a series of important papers that were adding more and more to knowledge of the phenomenon. Deubner was aware of the quality of data, in particular the high resolution, needed to verify the theoretical ideas advocated by Roger Ulrich and John Leibacher. In the early 1970s he therefore planned 'a series of new observations … designed to yield information on the existence of any fine structure … related to a modal character of the wave pattern'.[11] Like Roger and Ed, he too was searching for ridges.

The data that would give the first tantalising glimpse were collected in September 1974. Deubner scanned the Doppler shift along a part of the surface covering a length of about one-fifth of the solar diameter. The contours on the resulting k–ω plot indicated the presence of three, possibly four, ridges, spanning a horizontal wavelength range from roughly 6000 to 60,000 kilometres (Figure 4.4a). It seemed that Deubner had

---

[11] F.-L. Deubner, *Astronomy and Astrophysics*, 44, 1975, p. 371.

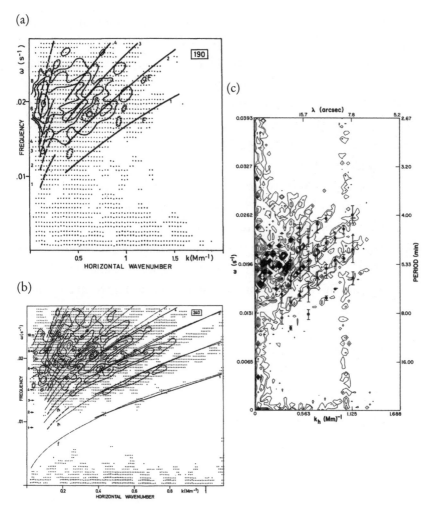

Figure 4.4. (a), (b) The k–ω diagrams from Franz Deubner's breakthrough paper of 1975. (a) The discovery diagram, made from data collected in September 1974. (b) The higher-quality diagram added in proof, made from data collected in the summer of 1975. Figures from Deubner, *Astronomy and Astrophysics*, 44, 1975, p. 371; reproduced with the kind permission of *Astronomy and Astrophysics*.

(c) A k–ω diagram made from data collected by Ed Rhodes, Roger Ulrich and George Simon in late 1974 and early 1975. The points, with error bars, show results from Deubner's observations. Figure from Rhodes, Ulrich and Simon, *Astrophysical Journal*, 218, 1977, p. 901; reproduced with the kind permission of the American Astronomical Society.

successfully verified the theory and in April 1975 he submitted his findings to the journal *Astronomy and Astrophysics*.[12] His discovery was left in no doubt whatsoever by a note added to the paper at the proof stage. This contained results of subsequent, and notably superior, observations made in June 1975. The new k–ω diagram showed of the order of ten beautiful ridges (Figure 4.4b). Deubner had found 'fundamental modes of sub-photospheric standing acoustic waves' – modes trapped beneath the surface of the Sun.

Independent confirmation was provided by the results of Ed and Roger, who, with George Simon, submitted their findings to the *Astrophysical Journal*[13] (Figure 4.4c). In the space of half a decade the field had evolved from being the study of something thought to be an atmospheric phenomenon to one of standing waves trapped within the solar interior. But the observational picture was still far from complete.

---

[12] Ibid.
[13] E.J. Rhodes, Jr, R.K. Ulrich and G.W. Simon, *Astrophysical Journal*, 218, 1977, p. 901.

5

GOING GLOBAL

The observational detection of ridges in the k–ω diagram was a crucial landmark in helioseismology. It was now clear the 5-minute oscillations were the visible manifestation of trapped, standing sound waves in the interior. However, the modes that had been observed engaged only a small fraction of the solar volume in oscillation – in the outer few thousand kilometres of the convection zone – and were still regarded as being a local phenomenon.

The possibility the 5-minute oscillations might include pulsations driving substantial fractions of the entire volume was raised in the early 1970s by Charles Wolff.[1] He computed frequencies from a model of the global Sun. In a paper published in 1973 he asked,[2] 'To what degree is the Sun pulsating at a considerable depth like a variable star and to what degree is its trembling merely a local response to convective overturning in its surface layers?' He seems to have favoured the 'variable star' outlook.

---

[1] In C.L. Wolff, *Astrophysical Journal*, 177, 1972, p. 87, Wolff presents a model of pulsations of the entire Sun.

[2] C.L. Wolff, *Solar Physics*, 32, 1973, p. 31. In this paper, Wolff weighs the evidence regarding the likelihood the horizontal length scale of the 5-minute oscillations may be somewhat larger than initially thought.

Several groups pursued observational strategies throughout the 1970s that would allow them to address directly this very question – groups based in Birmingham, Nice, the Crimea, Arizona and California. However, this was not always by design.

The Birmingham team, led by George Isaak, would discover the truly global nature of the oscillations. However, George and his colleagues had begun their work with a set of science goals completely unrelated to the Sun. It was only the idea of using sunlight to test their apparatus that led them by chance to the discovery the entire Sun pulsated as one unit in its radial, or breathing, modes of vibration.

George Isaak was born in Poland in 1933 to parents of German Lutheran extraction. He and his family emigrated to Australia while George was in his late teens. Several key events in his childhood and early career helped to lay the foundations of his helioseismology research.

George was fascinated by astronomy. He became a serious amateur observer of variable stars, like Cepheids, and a member of the Australian Branch of the American Association of Variable Star Observers. The brightness of these objects varies to such an extent that the pulsations – with their characteristic periods ranging from days to several weeks – can be discerned by visual observations with a small telescope. George spent many nights accumulating careful records on a variety of stars. But this remained very much an after-hours activity.

After taking his BSc and MSc degrees in physics at Melbourne University, and while dabbling in research on cosmic rays, George took up a full-time position at ICI Research Labs. There, during his spare time, he used the available resources to further his interest in yet another area, atomic and nuclear physics, and he built an atomic beam spectrometer from scratch. The concept of the instrument was novel to say the least. It provided a means to measure very precisely the positions of atomic spectral lines like the Fraunhofer lines of the Sun. As development of the instrument progressed, and it became clear this extremely novel design would actually work, and work well, George was allowed to concentrate full-time on the project. Details of the instrument were recorded in a submission to *Nature* in 1961. Then George began to ponder possible applications. These musings were brought into sharper focus by two research papers, one published in 1959, the other in 1961.

George was already aware of the former, written by Cocconi and Morrison, which speculated on the use of an atomic transition of the 21-centimetre-wavelength line of hydrogen as a means of interstellar communication. But it was when the second paper appeared, by Schwartz and Townes, that he really got excited. Instead of the hydrogen line they proposed the use of high-power wavelength lasers emitting in the visible range. A laser could be tuned to emit at a wavelength that would lie at the centre of a Fraunhofer absorption line in the spectrum of the parent star of the civilisation. If fired in the direction of an onlooking, distant observer, that observer would see a spike of laser radiation super-imposed on the profile. The properties of the spike could then be modulated to serve as a means of communication. George realised that, at least in principle, he had the apparatus at his disposal to scan for unusual features in stellar lines. Something else struck him too. It occurred that the atomic beam spectrometer could also be used to search for planetary companions of distant stars.

One might think the obvious way to find such a planet is to point a telescope and look, very carefully, for its light. However, planets are extremely faint because we only see them by the starlight they reflect. Technology has now moved on sufficiently to begin to make this direct approach a viable option. However, this was not the case when planet finding began. But all was not lost – it is possible to infer the presence of a planet by subtle, indirect means.

A planet tugs gravitationally at its parent star – just as the star pulls the planet. This means that the two bodies orbit at the same period around their common centre of mass. Take a familiar example. The Sun is about one thousand times more massive than the gas giant Jupiter. The size of orbit the Sun executes as a result of the pull of Jupiter is therefore about a thousand times smaller than the orbit of the planet. In fact, the radius of the Sun's orbit is just fractionally larger than the size of the Sun itself. The effect of the tug exerted by Jupiter is therefore to make the Sun appear to wobble gently backwards and forwards, on a timescale of just under 12 years. The velocity associated with this motion is, on average, about 13 metres per second (a sedate 50 kilometres per hour). Detect this motion – or rather the change in velocity (which is in principle easier to do) – and you can detect the planet.

George's atomic beam spectrometer was able to collect the sort of data needed to do this. That is because it could measure the position, in wavelength, of an absorption line formed in the atmosphere of the star. If the star were to wobble, so too would the line (the Doppler effect) and with it the signal recorded by the instrument.[3]

By 1969 George held a position in the Physics Department at the University of Birmingham. There, he had acquired sufficient seniority to begin to set these ideas in motion. His was an extremely ambitious pro-gramme, and so it was going to need darned good instrumentation. To get anywhere in planet detection demanded high sensitivity. The detec-tion of Jupiter-like planets would require that an instrument be able to monitor changes of a few metres per second against the backdrop of other possible variations of the star, the motion of the star with respect to the instrument on Earth, and noise.

Timescales were important. If the solar system were anything to go by, long-term monitoring would be needed to uncover signals with periods of the order of years. So, the instrument had to have good long-term stability. But from the signal-to-noise perspective collecting enough light from a potential target star was the key. The Sun dominates the sky. Move it to the distance of the nearest stars and its brightness drops by a factor of about 10^{12} (i.e. one million times one million) – and so, therefore, does the amount of light any instrument can collect.

---

[3] The wobble technique yielded up the first successful detections of extra-solar planets in the mid 1990s. There is now a substantial literature in this area. Other techniques are also now being applied. Careful observations and analyses should yield up detections of Earth-sized planets.

To alleviate the problem a large light-collecting area is needed. However, George was under no illusions about the prospects for securing long-term access at a large international telescope to observe the brightest stars in the sky! This precious time is usually reserved for the faintest objects, lying well beyond our galaxy, which cry out for a large light-collecting area. George therefore proposed that he and recently recruited PhD student Bill Brookes build their own mirror. The mirror did not require state-of-the-art optical quality but it did have to be big. What they were after was, in short, a large light bucket – a stellar flux collector.

To make the collector George proposed to make use of the well-known fact that the exposed surface of a spinning liquid in the laboratory takes the shape of a paraboloid. This is just the form needed to reflect parallel rays of light from a distant star through a sharp focus above the surface. A cheap and ingenious way of manufacturing such a mirror was to spin, at constant speed, a layer of liquid epoxy resin (essentially araldite glue) on a circular dish. The dish would be mounted on a low-friction air bearing to insulate the resin from the effects of vibration. The aim would be to keep the speed of rotation as constant as possible as the resin set hard to form the mirror.[4]

The production of a 2-metre mirror was set as the first goal. Unfortunately, many problems were encountered during testing and development. In particular, it proved hard to control the setting of the resin and this seemed to be the limiting factor in getting a sufficiently smooth surface. George and Bill therefore changed tack. They decided to concentrate on persuading the scientific community that Doppler shifts of absorption lines of stars could indeed be measured to levels of precision of a metre per second. To do this they intended to test the spectrometer on a star much closer to home – our star, the Sun. They decided to observe two closely spaced absorption lines in its spectrum, which are

---

[4] Bob Leighton – who we recall from Chapter 4 discovered the 5-minute oscillations in observations made with student Bob Noyes – went on to make a 1.6-metre epoxy mirror in the mid 1960s. This was used to perform infrared observations of very cool stars.

formed at visible wavelengths by sodium atoms. This would allow George and Bill to monitor the stability and noise in these lines and to check their suitability for planet finding and interstellar communications. The brightness of the star, and the resulting count rate in the instrument, would of course not be an issue.

Even though they were in a position to monitor the Doppler shifts on different parts of the solar surface they decided not to resolve the Sun into many patches or pixels. The project was, after all, a testing and proving ground for potential *stellar* observations, observations that would not have been able to resolve the surfaces of their targets into many patches because of the immense distances involved. The instrument was therefore designed to make 'Sun-as-a-star' observations, the recorded Doppler shifts of the sodium lines being an average from the signal collected from the whole of the visible surface. For the time being – in the end a gap that grew to some twenty years – mirror production was set to one side.[5]

George's team was bolstered in late 1969 by the addition of Bob van der Raay. Bob spent the first years of his life in New York. There, he and his parents had a brief, and unwelcome, flirtation with fame. In 1932, the son of pioneering aviator Charles Lindbergh was kidnapped from his home in Hopewell, New Jersey. The story attracted huge media attention and public interest. Very briefly during its course a young Bob was misidentified as the missing Lindbergh child.

Bob's late teenage years were spent in South Africa. Through university, and later after moving to the United Kingdom to take up a position at Birmingham, he began to specialise in high-energy physics, studying the interactions of particles under the high temperature conditions that prevailed in the very early universe.

---

[5] A 2-metre-diameter mirror was manufactured successfully in the early 1990s by the author – who was then pursuing this research as part of his PhD – and his supervisors, George and colleague David Bedford. We used the mirror to make observations of the giant star Arcturus. Alas, the mirror work was taken no further because the attention of the author was soon diverted from stellar to solar matters.

In 1969, the synchrotron particle accelerator at Birmingham was closed and several of the high-energy physicists had to transfer to the CERN particle physics centre in Switzerland. Bob did not want to move with his family and was therefore considering a change of tack so that he could stay in Birmingham. Over dinner with George and their wives Bob expressed a desire to do something 'fundamental'. It was then that George raised the possibility of being able to detect gravitational waves – from the changes they would give rise to in the Sun–Earth distance – with his spectrometer.

When charge is accelerated – like an electron in a television transmitter – it emits electromagnetic waves. Likewise, when masses are accelerated they are believed to emit gravitational waves. It is expected the universe is awash with gravitational radiation – but at an intensity that makes detection a challenging task, to say the least. As I write this book, experiments involving major international collaborations are thought to be in sight of the goal of detection. Back in 1970, George felt his apparatus offered a technique that was worth looking into further: could it be used to detect these effects through accurate and precise measurement of the velocity of the Sun with respect to the observer on Earth?[6] Bob certainly felt this was fundamental enough for him, and so was now on board.

Development of the instrument began, now with the triple-pronged goal of demonstrating feasibility for detecting planetary wobbles, uncovering unusual line features and detecting gravitational waves. The instrument was christened G-WAVE in honour of the most recently added objective. However, throughout this period George still had at the back of his mind the observations he had made of variable stars as a teenager and young man. Might the Sun exhibit similar fundamental and low-overtone oscillations?

Initially progress was slow, but steady. Partway through 1972 George made the decision to alter a fundamental part of the spectrometer design. The instrument was to work by comparing the frequency of a Fraunhofer

---

[6] George soon realised he was wrong to think this might be sensitive enough to yield a detection.

line of sodium, formed by atoms in the atmosphere of the Sun, with that of the same line formed by sodium atoms in the spectrometer itself. Motion of the Sun with respect to the observatory would Doppler-shift the solar and instrument lines with respect to one another.

In order to provide a stable wavelength reference in the instrument the original design contained an oven within which a small quantity of sodium metal was heated to about 180 °C. At this temperature it formed a vapour. The vapour was allowed to escape through a small hole in the oven, which collimated, or directed, the atoms to form a well-defined atomic beam. A beam of incoming light from the star was then directed to intercept the atomic beam, allowing the two to interact.

The small physical region where the two collided restricted the potential output signal, and so the decision was made to switch to an atomic vapour configuration. Here, the incoming sunlight was instead directed into the oven itself – now a small cell made of glass – within which the sodium vapour was contained. This gave a much larger cross-section for the sunlight to interact with the atoms in the vapour, boosting the potential output signal. At the same time it removed stability complications arising from that element of the instrument that had strongly directional (vector) properties – the atomic beam itself.

By 1973 the new design was working using sodium lamps in the laboratory as proxies of the Sun. That year another change was made as the team switched from using sodium to a line formed by potassium in the near infrared, which had several potential advantages. The potassium line lay further in wavelength from the potential corrupting influence of absorption lines formed by water vapour in the Earth's atmosphere. These could otherwise have added their unwanted Doppler shifts to the measured signal. Furthermore, potassium did not have to be heated to the high temperatures needed to vaporise sodium. This reduced the wear and tear on the oven in the instrument.

For reasons of practicality observations of the Sun began in Birmingham – to be more precise, on the roof of one of the buildings of the Physics Department – in the summer of 1973. At last the spectrometer was

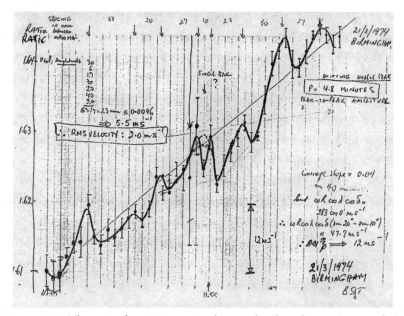

Figure 5.1. That page from George Isaak's notebook – observations made in Birmingham on 21 March 1974. The oscillation present on the main curve is the 5-minute signal. George's back-of-the-envelope calculations pepper the page, including, near the top right-hand side, his estimate of the period – 4.8 minutes. Image reproduced with the kind permission of G.R. Isaak.

receiving sunlight. After several tweaks and some tuning, a strong signal was being recorded whose signature bore that of the potassium atoms in the atmosphere of the Sun. Being the careful scientists they were, George and his team recorded, in fine detail, the observations in their logbooks. A page from one of these stands out immediately when scrutinised today (reproduced in Figure 5.1). In notes and pictures written by George, it is an account of observations made on 21 March 1974.

The basic observable is a measure of the Doppler shift between the potassium line on the Sun (average over the visible disc) and that in the instrument. This is proportional to the line-of-sight velocity between the Sun and the instrument. By far the largest component in the signal arises from the rotation of the Earth. In the morning, the rotation moves the observer toward the Sun. From the observer's perspective, it appears therefore as if the Sun is moving towards them. The Sun is 'blue-shifted' (in Doppler-speak) with respect to the instrument. At local noon, the Sun is at the meridian, as high in the sky as it will get. The direction of its motion is then all at right angles to the Sun–Earth direction and the line-of-sight component from the spin is zero. In the afternoon, the rotation now appears to carry the Sun away from the observer. The signal from the Sun appears red-shifted.

We would therefore expect the velocity of the Sun with respect to the observer to increase steadily throughout the day, going from negative (in the morning, by convention), through zero at noon, to positive in the afternoon. The shape of the resulting curve follows from understanding the motion involved. The line-of-sight velocity measured is the component, in a particular direction (here the Sun–observer line), of motion in a circle. So, the curve is just a sine wave whose amplitude is a few hundred metres per second.

Rotation of the Earth is not the only large-scale contribution to the signal. The orbit the Earth follows around the Sun is not circular but elliptical; the distance at closest approach is about three per cent smaller than when it is farthest away. This means the component of velocity parallel to the Sun–observer line changes by a few hundred metres per second. However, this takes place over the course of a year, rather than a day. The third large component is the gravitational red shift. This positive-velocity offset arises because the photons making up the beam of light detected by the instrument must first climb out of the gravitational potential well formed by the Sun. This costs energy. Since the energy of each photon is inversely proportional to its wavelength, photons must pay to escape by having their wavelengths increased (or red-shifted). The size of this velocity signal is about 630 metres per second.

The 21 March 1974 data plotted by George show results collected over a few hours. The slow but steady increase in velocity caused by the rotation of the Earth is clearly seen. However, superimposed on this is what appears to be an oscillating signal with a period well under an hour.

Parts of the page surrounding the plot were peppered with carefully worked-through back of the envelope calculations, which appear to have been done at a furious pace. First, George performed a calculation to convert the raw, measured signal into absolute units of metres per second. This allowed him to establish that the steady increase in the size of the signal was indeed attributable to the Earth's rotation. He could now estimate the velocity amplitude of the rapid 'oscillation'. This turned out to be about a metre per second. Finally, it was a simple matter to estimate an average period for the oscillation from the handful of complete cycles recorded on the plot. George did the calculation and wrote and ringed the result on the right-hand side of the page – 4.8 minutes.

Without realising it at the time George and his colleagues were observing 'whole Sun', globally coherent 5-minute modes, oscillations for which the surface was moving in concert on huge global length scales, each part hooked into what the other parts were doing. The modes seen were those formed by sound waves that penetrated the deep interior and core. Some (the radial or breathing modes) engaged the entire volume in oscillation. This was in marked contrast to those modes whose ridges would soon be uncovered by Deubner and Rhodes et al., which penetrated only the outer layers of the convection zone. So, why was it these two classes of observation were to turn up such different modes?

The answer lay in the way each observing strategy treated the visible disc of the Sun. George and his colleagues had opted for a Sun-as-a-star approach for the reasons already explained. This meant they averaged signal over the Sun's visible disc; which meant high-wavenumber modes left a negligible signature in their data.

High-wavenumber modes have short horizontal wavelengths. The shorter the wavelength the smaller also is the distance between consecutive reflections on the solar surface. The three-dimensional patterns of oscillation these modes give rise to have numerous, adjacent small patches marking where the surface is moving towards or away from the observer (a 'chequerboard'). The tendency will be for these negative and positive-velocity shifts to cancel one another when the cumulative effect

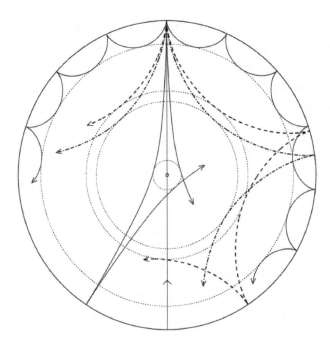

Figure 5.2. Cross-section of the solar interior, showing several paths followed by sound waves of different wavenumber, including some low-wavenumber waves that penetrate to the core (the two most deeply penetrating shown). A dotted circle traces out the lower turning point for each wave. Modes established by the low-wavenumber waves are visible in Sun-as-a-star data (see text). The spherical harmonic angular degree can be used to identify modes instead of wavenumber (recall discussion at end of Chapter 3; and see later in this chapter). The shallowest-penetrating wave here has an angular degree of $l = 75$. A mode formed by it would not be detectible in Sun-as-a-star data. In contrast, the following two would: the straight ray passing through the centre has $l = 0$ (it is radial), and the next most deeply penetrating wave has $l = 2$. Image reproduced with the kind permission of J. Christensen-Dalsgaard.

is measured across the disc. The net signal that remains is then so small as to be dwarfed in comparison with that left by low-wavenumber modes.

Low-wavenumber modes have large horizontal wavelengths and give very simple patterns of oscillation across the surface (Figure 5.2).

The signals from many of these modes do not cancel to zero[7] when averaged over the disc, and so dominate any Sun-as-a-star view. What is more, with a low wavenumber comes a greater depth of penetration, which is why the Birmingham data revealed the strong signature of core-penetrating modes. In contrast, Deubner and Rhodes et al. had imaged the visible disc in sufficiently small patches to allow them to track the high-wavenumber, shallow-penetrating modes.

Much later George was to admit that the thought the Sun might oscillate with such a rapid 5-minute period just had not occurred to him when these first data were collected. In this view he was guided by prejudice from his earlier variable-star observations. These led him to believe that were the Sun to oscillate at all, it would do so in its fundamental and low-overtone modes of oscillation like a Cepheid variable. This leads to an expected period for the Sun of the order of an hour, rather than several minutes. As George was sometimes fond of saying, 'you often see only what you want to see'.

The explanation the Birmingham team favoured at the time was that they were seeing the effects of granules at the top of the convection zone. The timescale for this motion had roughly the correct period. Furthermore, George's ubiquitous back-of-the-envelope calculations suggested that, when averaged over the whole of the visible disc, the predicted amplitude of the granular signal was at about the metre-per-second level recorded by the Birmingham instrument. But George, Bill and Bob were not the only folk making Sun-as-a-star observations – Eric Fossat and his colleagues in Nice were reaching the very same conclusions regarding the likely origin of the 5-minute signal in their data.

It is spring 1968. Eric Fossat had just begun his PhD under the supervision of François Roddier. In one of their first meetings Roddier opened

[7] Some of the rotationally split components in each mode multiplet are missing because the net average of their Doppler signal over the visible disc is so small. The term 'visible disc' is key here. Earth-bound observations view the Sun from a perspective where the solar rotation axis is in the plane of the sky. Were we instead offered a view that looked directly down on one of the solar poles, those modes that were missing from the Earth-bound Sun-as-a-star observations would instead stand proud, and vice versa. All the mode components are in principle accessible from an angle of view intermediate between the extreme cases.

his desk to extract a small, delicate vial-like container made of glass. Held within it was a small quantity of sodium metal. 'With this small cell', announced Roddier, 'we can build an instrument to study the Sun.' What Roddier had in mind was a spectrometer capable of making very high-resolution observations of the solar granulation. The solar 5-minute oscillations were not part of this initial plan.

The French system then required that graduate research students submit two theses – a first that would take two years and a second advanced thesis that would usually take an additional five. For his first thesis, Eric had the objective of developing an instrument to observe the granular signal. Initial effort was devoted to building a test-bed instrument in the laboratory, which shared its basic working philosophy with the Birmingham instrument. By 1970 a telescope-sized unit was ready for use at the Rome Observatory. It was able to record a strong signal from patches of the Sun. Eric had thereby demonstrated the observational concept was a good one. However, sky conditions in Rome were not good enough to allow the fine detail of the granulation to be seen. But another feature was showing up in the data – the 5-minute signal.

For his second thesis, Eric therefore shifted the emphasis of his work towards a detailed, systematic study of the properties of the 5-minute signal. The first observations dedicated to this were made by Eric, and Gilbert Ricort, at the Pic-du-Midi Observatory, in the French Pyrenees, in 1971. Circular apertures were used to pick out portions of the Sun to observe, which ranged in size from about one to eighteen per cent of the solar diameter. What Eric found was something of a surprise. Even for the largest apertures the 5-minute signal remained clear and strong, in spite of the poor observing conditions. This suggested the phenomenon was coherent on length scales that extended over a distance of many tens of thousands of kilometres or more on the solar surface. This was certainly a larger number than had previously been entertained – the phenomenon was still thought to be a local one, with a length scale of perhaps a few thousand kilometres at most.

By 1973, some of the more recent 5-minute data – in particular Eric's set – had begun to raise serious doubts about a local interpretation for the signal. Charles Wolff discussed this issue in his 1973 paper in which he sought to prod his colleagues into action by pointing out what he felt had been 'reluctance in the literature to try to understand the apparently great inconsistency between the measurements of large scales for the 5-min oscillations and the earlier reports of small scales'.[8]

Between 1972 and 1974 Eric threw himself into a complete overhaul of the solar telescope at the Nice Observatory. Only by having dedicated access to a telescope did he feel he could make significant headway. Once they were ready, Eric and his colleagues extended the observations to include looking at the entire visible disc of the Sun in one go. And as in Birmingham, their Sun-as-a-star data showed a strong, healthy signal centred on 5 minutes.

Deubner and Rhodes et al. had by then provided observational confirmation for the existence of high-wavenumber modes in the interior. However, the uncovered modes only tickled the outer layers and certainly could not explain the Birmingham and Nice observations. One possibility was that a different set of much longer horizontal-wavelength modes was being seen – the whole-Sun modes Charles Wolff had alluded to back in 1972 and 1973. The Birmingham and Nice groups both published data showing the 5-minute bump of power standing proud, but for now no one made the global-mode connection. Hindsight is a wonderful thing. Nevertheless, there were sound reasons for supposing the Sun-as-a-star signal was dominated by granulation covering the surface, because Eric and George both reached this same conclusion.

One must remember that the 5-minute signal was then something of a distraction for the Birmingham team. The original objectives of the team had called for a demonstration of excellent instrumental sensitivity and stability on much longer timescales. It should therefore come as no surprise that it was towards these goals that most of the effort had been

[8] C.L. Wolff, *Solar Physics*, 32, 1973, p. 31.

directed. Only a few months after uncovering the 5-minute signal in the data, Birmingham and another group based at the Crimean Astrophysical Observatory both appeared to have hit the jackpot – evidence for an oscillation of the entire Sun with a much longer period.

The group in the Crimea, led by Andrei Severny, with colleagues Valery Kotov and Teodor Tsap, had previously been making observations of magnetic fields in the photosphere. In 1974 they modified their magneto-graph to make measurements of the Doppler shift of the solar surface. These observations were made in a slightly different way to the Birmingham team's. The Crimean instrument measured the difference between two Doppler shifts, each an average over different parts of the disc – one a circular region at the centre, the other a circular annulus on the outer rim.

In a series of observations made over the summer and autumn of 1974, Severny, Kotov and Tsap found what appeared to be a strong, significant oscillation with a period of about 160 minutes and a Doppler velocity amplitude of just over 2 metres per second. The fact the oscillation showed up prominently in their data meant that, if it really was solar in origin, it had to be global in nature, with a huge horizontal wavelength. Severny interpreted it as being the fundamental radial breathing mode of oscillation of the Sun. Alas, this interpretation was to prove incorrect.

Meanwhile, George Isaak had spent the first few months of 1974 searching for a better site from which to make his observations. Those with even a passing knowledge of English weather will recognise that Birmingham was a location rather ill-suited to the task at hand. Over Easter 1974 George visited the Pic-du-Midi Observatory and come the summer Bob van der Raay and Bill Brookes were on site to set up an instrument. Bill stayed for several months collecting data. He also witnessed some extreme weather. During one freak episode wind speeds on the observing plateau peaked at over 200 kilometres per hour, sweeping away one of the large observatory domes (though fortunately for him not the Birmingham instrument).

In the early months of 1975 these data were analysed carefully. Some of the early, preparatory phases were done collectively. The key later parts, which might or might not have uncovered a solar signal, were done individually. This independent approach would offer some protection against any interesting result being merely an artefact of the analysis. In the modern era of helioseismology independent checks of this nature – more grandly referred to as the 'use of independent data pipelines' – can play an important role in verifying the solar origin, or otherwise, of subtle features. Results from the same data can differ for a host of reasons: for example, slight differences in the approach used to arrive at the result, or the impact of the way in which the approach is coded in the software. The more complicated the problem, the more varied are the number of possible approaches. It is then the use of different pipelines becomes important.

The mid 1970s were the days of paper-tape programming. Bob van der Raay was usually the quickest off the mark in the Birmingham team to analyse data, and so it proved when, in April 1975, he came to see George with the results of his Pic-du-Midi analysis. These showed a strong 160-minute oscillation. George had just completed writing a new grant application for funding to continue the work. The application was due at the Science Research Council headquarters that day. With no time left to make any substantial changes, but mindful they must include this exciting result, George put a handwritten note next to the part that said to date no oscillations had been uncovered – '*BUT*, see enclosed graph'.

The Crimean and Birmingham teams had uncovered the same signal without being aware of the other's work. George would only learn of the Crimean result when he attended a meeting in London in September 1975. In the meantime both groups submitted their results independently to *Nature*[9] and continued to chase the 160-minute signal. From 1976 another team, based at Stanford University in California,

---

[9] Both papers appeared in the same edition, along with the seminal theoretical paper, by Jørgen Christensen-Dalsgaard and Douglas Gough, referred to in Chapter 1. A.B. Severny, V.A. Kotov and T.T. Tsap, *Nature*, 259, 1976, p. 87; J.R. Brookes, G.R. Isaak and H.B. van der Raay, *Nature*, 259, 1976, p. 92.

began seeing it too. With three independent groups observing the same phenomenon, surely this had to be a real solar oscillation?

The proposal for what would become the Stanford solar telescope included as its main aim the study of large-scale magnetic fields covering the solar surface, and the 'fast' solar wind. The Sun loses matter continuously via the solar wind. Lines of magnetic field and the solar plasma tend to be tied together, and in the outer corona matter streams out into the solar system along 'open' field lines. The outflow can reach velocities of a few hundred kilometres per second. The flow speed is about twice as fast near the poles, in what are called 'coronal holes', than in the vicinity of the solar equator. Coronal holes are where the field lines extend out to very large distances above the surface, well into the interplanetary medium. In the early 1970s, when the proposal was written, there was some speculation the fast wind might in some way be energised by acoustic waves. The Stanford team therefore had an interest in studying the 5-minute acoustic waves.

Phil Scherrer, who with John Wilcox (and latterly graduate student Phil Dittmer) had begun to make observations of the 5-minute oscillations, remembers travelling to the Crimean Astrophysical Observatory to observe the work of Severny and colleagues. The Stanford telescope was already up and running. When it became clear that the centre-to-rim difference method used by Severny could give the Stanford set-up better sensitivity, Phil set about contacting the Stanford team from the Crimea.

This was not easy. Telephone calls took several hours to organise and communications were more often than not achieved via a combination of telegrams and letters. When he finally spoke to Stanford by phone, Phil told them to install a similar system with an annulus to let through light from a central portion of the disc which was 'half the size of the Sun'. Because of the poor quality of the line and the high likelihood it might go down during the call, Phil did not have time to be more specific. That is why the Stanford system ended up being slightly different. His colleagues interpreted 'half the size' to mean fifty per cent by radius. The central portion of the Crimean system covered just less than fifty per cent – by area.

Although the initial Stanford results failed to detect the 160-minute oscillation at the level reported by Severny, later analysis of results covering the period 1976 to 1978 tended to confirm the Crimean findings. So, as time wound on into the later 1970s, there seemed to be indisputable evidence from the observations that the Sun was undergoing oscillations that most likely engaged in motion a large fraction of its entire bulk. In spite of this, the identification of this 160-minute oscillation as the basic resonant mode of the whole Sun remained controversial. Finding a plausible origin for the signal was still a problem that refused to go away – the signal persisted in the data, at least for now. (We shall return to the 160-minute story in the next chapter.) In the meantime the focus of the Birmingham group was to shift again, but this time *back to* the 5-minute region as further analyses of the 5-minute feature began to suggest there might, after all, be more to it than previously thought.

Clive Mcleod joined the Birmingham group formally in the mid 1970s. He had already been helping the group on and off and his expertise in electronics and computing had already been put to good use. He designed and built all the principal electronics of the group's instrumentation up to his retirement in the 1990s. After getting up to speed very quickly, Clive expressed an interest in becoming involved in the analysis of data. But what could he do? The answer George and Bob gave reflected their continued belief that the 5-minute signal probably had nothing dramatic to offer. 'This isn't terribly exciting,' admitted George, 'but you could take a closer look at the 5-minute bump.' George still felt it was probably the signature of convective granulation. However, the characteristics of the bump needed to be properly measured, and there might also be the possibility of these changing with the Sun's 11-year cycle of activity.

In the end it was the combination of this extra manpower and two other factors that led the Birmingham team to look with a more inquisitive eye – the extra factors being the accumulation of greater numbers of data and the acquisition of more powerful computers with which to analyse them. Bob – who at the time had the fastest and most powerful computer in the group – and George – whose machine took a little longer to crunch the numbers – began to make spectra from longer and longer stretches of data, which showed the power in their Doppler measures as a function of frequency. These longer datasets brought with them an improvement in frequency resolution.

The data available to any observer are finite in extent. This sets a limit to how finely they can be sampled (or their spectrum divided) in frequency. Power is assessed on a discrete grid of 'bins'. The spacing of the grid (the bin width) is set by the inverse of the length in time of the observations. Put simply, the more data you can accumulate the better able you will be to determine the frequency of any feature that is present. This – and the way the intrinsic resolution is fixed by length – follows in the spirit of Heisenberg's uncertainty principle.

The 5-minute signal in the Birmingham data left its mark in the form of a strong bump of power in the power spectrum, centred on a frequency of 3 milli-Hertz and spanning a range of roughly 2½ to 4 milli-Hertz (a range of just under an octave). This also happened to match quite well the frequency range over which high-wavenumber modes had been observed. When they looked at stretched-out plots showing this part of the spectrum – as made by various subsets of their data – George, Clive and Bob were moved to comment that the bump took on the appearance of something altogether more interesting: a collection of narrow peaks. George, in particular, remained sceptical whether the effect was real (by which, in this context, we mean consistent and repeatable). Given the resolution in frequency then achievable, he had good reason to be. But this potentially intriguing signature meant there was now even more of a reason to get to the bottom of the properties of the 5-minute feature.

A more reliable flow of high-quality data had come on tap when, in the summer of 1975, the Birmingham team moved its instrument to the Observatorio del Teide at Izaña, on the island of Tenerife. This marked the beginning of a successful, long-standing collaboration with the host institute, the Instituto de Astrofísica de Canarias, which continues today. George and Bob also took on a PhD student from the recently

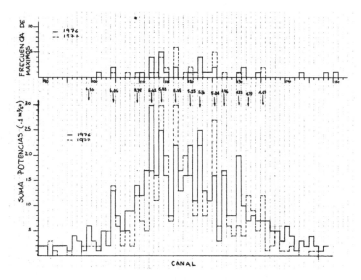

Figure 5.3. An original plot from the PhD thesis of Teo Roca-Cortés, revealing the tell-tale picket-fence pattern of peaks from the low-degree modes. It shows the average power spectrum (power against frequency, axes labelled in Spanish) for data collected by the Birmingham team in 1976 (solid line) and 1977 (dotted line). Teo has annotated the periods of identified peaks, which run from just over 4½ to 6½ minutes. Image reproduced with the kind permission of T. Roca-Cortés.

established institute to work with them. Teo Roca-Cortés had as the main goals of his project to look not only at the still present 160-minute oscillation but also (and as it turned out crucially) at the 5-minute signal.

Teo's presence at Izaña meant the observing runs of 1976 and 1977 could be extended. He divided his time between Tenerife and working on the data in Birmingham. By 1978 he was well and truly focused on the 5-minute region. Teo was able to make power spectra from the many individual days of data accumulated during each of the 1976 and 1977 campaigns. He then made averages of these, one for each season. It soon became apparent that both average spectra were showing evidence of a regular grid, or picket-fence-like pattern, of closely spaced peaks. In every third bin or so in frequency, in the 2½–4-milli-Hertz range, Teo found a strong concentration of power well above the background noise (Figure 5.3). The mean spacing of these peaks turned out to be just under

68 millionths of a Hertz (68 micro-Hertz). Once he had convinced himself the pattern was genuine, Teo put together a report for George to read. George was extremely excited by the result and did some independent checks of his own. This left him convinced the spacing was real signal in the data. At first, Bob van der Raay was sceptical. However, after beginning his own full analysis of new data, collected over the summer of 1978, Bob began to waver. By the early summer of 1979 Bob was also convinced.

The result was significant indeed, for it stood on its head the notion of the 5-minute signal being local and confined to the outer layers of the Sun. Teo had uncovered very large, missing pieces of the 5-minute jigsaw puzzle – the fact the peaks were the signatures of the global modes of oscillation of the entire Sun – and a submission by the Birmingham team (Figure 5.4) to *Nature* marked this discovery.[10] The spectrum of a simple, one-dimensional pipe has a fundamental mode and higher overtones, which are spaced evenly in a harmonic series in frequency. This is a regular picket-fence pattern of modes. George and his colleagues realised they were seeing the same thing on the Sun – overtones of resonant modes of oscillation.

The fact the Birmingham team was making Sun-as-a-star observations meant that, first, the modes had to be global in nature and, second, the modes could only be those with the longest possible horizontal wavelengths whose waves happened to penetrate the deep interior. The simplest class seen were the purely radial, breathing variety formed by sound waves that travel right to the centre of the Sun – quite literally 'whole-Sun' modes. Were several to be excited to observable levels, they would have been expected to contribute their own picket fence of overtones to the spectrum.

Unlike the idealised one-dimensional pipe, the Sun supports families of modes of different horizontal wavenumber (like the transverse modes

[10] A. Claverie, G.R. Isaak, C.P. McLeod, H.B. van der Raay and T. Roca-Cortés, *Nature*, 282, 1979, p. 591.

in real three-dimensional cylindrical or conical pipes). As well as the aforementioned breathing modes there are a plethora of non-radial modes (Chapter 3). First come the so-called dipole modes. The pattern of displacement they give rise to at the surface makes opposing hemispheres wobble out of phase with one another. Next, in order of increasing complexity – and wavenumber – come the quadrupole and octupole families. Through this progression the depth of penetration of the modes gets shallower. However, the majority of overtones of the four types listed, which dominate the Sun-as-a-star signal, penetrate the core. At higher wavenumber the Sun-as-a-star mode visibility falls off quite rapidly; much higher still and the domain occupied by the modes observed by Deubner and Rhodes et al. is reached, with their shallow cavities.

Rather than ascribing the horizontal scale a wavenumber it was more natural to use the notation of spherical harmonic functions to describe the whole-Sun modes seen by the Birmingham team. From Chapter 3, the length scale of the oscillation pattern is then described by the angular degree, l. The higher the l, the larger is the wavenumber.[11] The radial modes have $l = 0$, the dipole $l = 1$, the quadrupole $l = 2$ and the octupole $l = 3$; these are the 'low-degree' modes.

With several overtones excited to observable levels one would have expected four sets of prominent picket fences in the Sun-as-a-star spectrum,[12] one for each angular degree (or wavenumber) seen clearly in the data. Though not precisely harmonic in character, the frequency spacings turned out to be fairly uniform in the higher overtones.

High overtones of the low-degree modes certainly seemed to be the ones uncovered by Teo. The fundamental radial mode was expected to have a period of about an hour, and the periods of the modes seen in the

[11] From the 1980s onwards it became more common practice to talk in terms of high-degree modes in an l–v diagram, rather than high-wavenumber modes in a k–ω diagram. Although I admit that chopping and changing between wavenumber and angular degree may cause some confusion it more accurately reflects the evolution of terminology in the field.

[12] The $l = 4$ and $l = 5$ modes can also be seen in the more recent Sun-as-a-star data. The early, short-dataset spectra were of insufficient quality to uncover these modes.

5-minute region were of course substantially shorter. The periods were in fact so short that the strongest mode could be identified as being around the twentieth overtone of the fundamental. The Birmingham team made a tentative stab at identifying which overtones were which, using as a guide the theoretical frequencies from a whole-Sun model computed by Icko Iben and John Mahaffy. Although the assignment they proposed was not quite right – the paper claimed no more, and it would take a little while for the field to properly identify the modes – it was nevertheless clear that about ten overtones of each angular degree were prominent in the data. But how did the four families of overtones – those for each of l = 0, 1, 2 and 3 – fit into the pattern of peaks seen?

The spacing in frequency between successive high overtones of the fundamental radial mode turns out to be about 136 micro-Hertz. This corresponds to the inverse of the time it takes for an acoustic wave to travel from the Sun's surface to the centre and back again, a round-trip of approximately 123 minutes. It is the same measure given by the spacing of overtones of a simple pipe. There, the round-trip is from one end of the pipe to the other and back again. The overtone spacings of the other modes in the Sun-as-a-star data are of a similar size. The arrangement of the resulting four patterns of overtones then depends on the placement in frequency of one family with respect to another. The result is certainly not intuitive.

It causes radial (l = 0) modes to lie next to quadrupole (l = 2) modes (typically a snug 10 micro-Hertz apart), and dipole (l = 1) modes to pair off with octupole (l = 3) modes (on average about 15 micro-Hertz apart). Different classes of pair lie about 68 micro-Hertz from one another. With insufficient resolution to separate out the modes in a pair, an observed spectrum will take on the appearance of a picket fence with a 68-micro-Hertz spacing. Which is precisely what Teo and the Birmingham team had seen.

In the immediate aftermath of the Birmingham team's discovery the field of helioseismology was faced with a large gap in the mode data, between

Figure 5.4. Corduroy to the fore! Members of the Birmingham team, circa 1979. From left to right: George Isaak, Bob van der Raay, Teo Roca-Cortés, Andre Claverie and Clive McLeod. Picture courtesy of G.R. Isaak.

the low and the high-degree modes. New observations would be required to bridge this gap, a story that awaits our next chapter. Of more pressing concern for the low-degree range was the need to separate observationally the even ($l = 0$ and $l = 2$) and odd ($l = 1$ and $l = 3$) pairs making up the Sun-as-a-star spectrum. The data required to do this were soon available, courtesy of Eric Fossat.

When we last caught up with Eric he had been looking at the 5-minute signal through different-sized apertures (including Sun-as-a-star observations) in the belief that what he was seeing was the signature of granulation, not another set of modes. This phase of study, and his second thesis, was completed in 1975. He was joined shortly after in Nice by Gérard Grec, who set about updating their sodium-based spectrometer to turn it into what was, in Eric's words, a 'modern instrument for the time'. For one thing this meant dispensing with paper-tape recording.

Eric determined they should now use this new instrument to look for potentially more interesting signals with longer periods (i.e. lower frequencies). There had been suggestions in some of the previous Birmingham and Nice data of tantalising signatures close to an hour. There was also the 160-minute oscillation. Eric decided to concentrate on testing the significance of features uncovered in the 40–60-minute range. Furthermore, he wanted to see if they might be correlated with signatures of activity like solar flares.

The data collected for this led to a paper, published in 1977, which is worth our spending a little time over. For one thing it actually shows evidence for the ~68 micro-Hertz pattern. Hindsight again! Eric and Gérard, for what were at the time sensible reasons, marked this out as an artefact of the stop-start nature of their observations. This appreciation of the problems gaps in coverage could create was a strong motivation for Eric to find a way to collect stretches of data spanning several days with a minimum of interruption. Later, it turned out this was precisely what was needed to resolve the individual modes in the Sun-as-a-star spectrum.

Eric and Gérard initially sought to improve the frequency resolution by stringing together observations made over consecutive days with their telescope in Nice. Provided it was sunny, each day of data comprised about four hours of data in the morning, followed by a gap – when the telescope had to be rotated on its mount – and a further four more hours in the afternoon. By making a power spectrum from a set of total length 4½ days they were able to improve the frequency resolution by more than a factor of twenty over what could have been achieved by looking at the 4-hour pieces individually. However, because they only had about 8 out of a possible 24 hours of data on each day, artefacts were introduced into the spectrum.

Even though the focus of their gaze was set firmly on the long-period, low-frequency part, Eric and Gérard included in the 1977 paper a plot of the spectrum rendered on a scale fine enough to reveal in detail the 5-minute bump. This undoubtedly shows the spacings of the pairs of 5-minute modes.

They had of course spotted this. But they felt it was an artefact of their gap-ridden data. We recall from above that the pattern of observation went something like: 4 hours, short gap, 4 hours, long gap; on to a new day and the same again. A 4-hour period is equivalent to a frequency of 69 micro-Hertz. Given the still modest resolution of their spectrum this would have looked pretty much the same as something with a 68-micro-Hertz spacing, and so seemed to explain what had been seen.

Eric was more determined than ever to improve the quality of his data. This meant reducing the gaps and getting better resolution. By 1977 he had moved on from Nice to a postdoctoral research fellowship in Boulder, Colorado. While there he spoke at length with Bernard Jackson, who had wintered over in Antarctica in 1975–6 making measurements of water vapour content and the observing quality of the sky. Might it be possible, Eric wondered, to make extended observations of the Sun with his apparatus from a scientific station in Antarctica?

Provided weather conditions were favourable this would allow continuous observations to be made spanning many days. By the spring of 1978 Eric's initiative had paid off after further conversations with Martin Pomerantz. Pomerantz had been a regular visitor to the Antarctic and had access to the Bartol Research Foundation observatory located only a few hundred metres from the geographical South Pole. Preparations for an observing run soon began.

Eric and his colleagues arrived in Antarctica on 23 November 1979. They brought with them the constituent elements of their instrument, whose final construction and integration they intended to perform on site. Temperatures outside fell to –35 °C, conditions all had to bear on their daily traversal from the main station to the telescope and instrument site, some seven kilometres away. Initially, this was done within the slightly warmer confines of a vehicle. However, repeated breakdowns meant the team had to resort to cross-country skiing. Eric suffered from frozen, but not frostbitten, feet.

The decision to locate the instrument some distance from the main site – to remove it from the pollution of the station – was a deliberately conservative one. Members of the team may have felt that they had erred too much on the side of caution as their strenuous cross-country trips mounted up. Nevertheless, the skies there were wonderfully crisp and clear and by Christmas the team was ready.

Eric recalls how the local circumstances demanded simple and efficient solutions be implemented when problems cropped up. Ingenuity was very much to the fore. Even though by the end of the year temperatures were still hovering around –20 to –25 °C, the glass cell containing the essential sodium vapour kept overheating when the wind died down. The remedy? To connect to the large magnet within which the oven and cell were housed a copper strip of sufficient length to allow one end to hang freely from the back of the instrument. This carried away the excess heat. The team also found that stubborn light leaks could be plugged by wrapping the offending parts of the instrument in cooking foil.

The observing run, which would generate the 'South Pole spectrum', began at 10 p.m. local time on New Year's Eve 1979. (The scientist is a rare, dedicated breed.) It lasted 6 days; just one short 2-hour stretch was obscured by cloud. The instrument worked well. Eric and his colleagues knew they had something that at the time was unique – an almost continuous set of multi-day observations.

Today, it is possible to turn raw data into a spectrum in just a few seconds. As I write, I am looking at spectra that were made from data collected this very day by two sites of the present-day Birmingham-run network, in Australia and South Africa. Things were not so rapid in the late 1970s. The team had first to contend with its temporary location at the South Pole. As the observing run ended, team members were several thousand miles away from their computers in Nice. Following the return of the scientists to Nice, in the early spring of 1980, the magnetic tapes on which the raw data were stored had first to be transferred to a computer, which was still fed with innumerable punched cards as input. The cards had to be punched out carefully – no hanging chads allowed here – by another machine and then loaded in sequence. Lots of intermediate steps were therefore required before a final result was

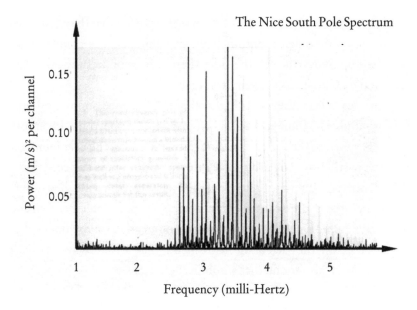

Figure 5.5. The fine structure of the whole-Sun spectrum is revealed in all its glory by observations made at the South Pole by Eric Fossat, Gérard Grec and Martin Pomerantz (Figure 5.6). Adapted from figure in Fossat, Grec and Pomerantz, *Nature*, 288, 1980, p. 541; reproduced with the kind permission of the Nature Publishing Group.

forthcoming. When it came (Eric remembers clearly Gérard arriving with the spectrum) the result was worth the wait. Together with theoretician Jean Paul Zahn they pored over the plot – it was now possible to see the expected range of individual modes (Figure 5.5).[13]

It had taken two decades, and the ingenuity and hard work of many individuals, but helioseismology had got there. Those first intriguing observations of Bob Leighton and his colleagues, and the full range of 5-minute oscillations supported by the Sun, could now be explained.

---

[13] The South Pole spectrum was published in G. Grec, E. Fossat and M. Pomerantz, *Nature*, 288, 1980, p. 541.

Figure 5.6. At the South Pole, Christmas 1979 – from left to right: Eric Fossat, Martin Pomerantz, Lyman Page (student of Martin) and Gérard Grec. Picture courtesy of E. Fossat.

6

EARLY SUCCESSES

The first attempts to put the scientific potential of the solar oscillations on a firm footing were made at the Institute of Astronomy, in Cambridge, in the mid 1970s. It is from the perspective of the two scientists who did this work – Douglas Gough and his then PhD student, Jørgen Christensen-Dalsgaard – that we begin our next chapter and consider the science that came out of the first observations of the modes.

Looking back, both Douglas and Jørgen regard 1975 as a key time for helioseismology in as much as it marked the real beginnings of a concerted effort to think seriously about how oscillation data might be used to infer properties of the solar interior. In a recent review of the field, Jørgen has coined 1975 the 'annus mirabilis' of helioseismology. At the time the field was not known by this name. Douglas would soon give it its Greek appellation, and both scientists would be at the forefront of the theoretical advancement of helioseismology over the years that followed.

Douglas had not always intended to pursue a research career in an astronomical subject. As he approached the end of his undergraduate degree in Cambridge the only thing he knew for sure was that he wanted to enter an area in which he could make a real impact. Why not, he thought, opt for a brand new topic? A topic so new that it would be crying out for bright folk to make their mark? The field he initially thought seriously about was theoretical chemistry. In order to make a more informed decision Douglas began attending chemistry lectures that he thought might be relevant to his ideas of treating chemistry like physics. Asked today to remember the subject matter, Douglas recalls clearly some lectures given in crystallography which, alas, sounded the death knell of the whole idea. Suffice to say they were a little dry for his taste.

Still determined to pursue a career in academic research, but now needing a subject within which to apply his considerable talents, Douglas turned his thoughts back to physics. In order to make a decision he then applied the following test: Who, he pondered, looked happy in the physics department in Cambridge? Who was really enjoying what they were doing? The firm conclusion he reached was that the astronomers and astrophysicists seemed a jolly, hard-working bunch.

Donald Lynden-Bell was the first of their number from whom he sought direct advice. Douglas was now thinking of relativity as a possible research area for his PhD. He asked Donald whether this would be a good choice. Donald's response was 'Do you think you will be successful?' Somewhat unsure Douglas replied, 'I don't know,' to which Donald retorted, 'Well, don't do it.' The tongue-in-cheek advice Donald left Douglas with was that the secret of success in research is to have an over-inflated opinion of your own ability.

Next, Douglas sought out Fred Hoyle. Douglas was fascinated by Hoyle and now hoped the eminent scientist would agree to supervise him for a PhD on theoretical aspects of cosmology. Hoyle's response was an immediate and emphatic 'no'. A slightly overawed Douglas plucked up the courage to ask why, to which he received the following answer: 'The trouble with cosmology' – and this was true then – 'is that there are essentially no observations to check the theory. So you cannot be proved wrong.' He then suggested Douglas choose an astrophysical topic for his PhD and said that if, after finishing, he still wanted to pursue research in cosmology Fred would do all he could to help.

Douglas thought this a fair answer. But he still sought a second opinion from another member of the department. Their response was 'Yes, go for cosmology.' When asked why, they replied, 'Oh, there are essentially no observations in cosmology, so you can't be proved wrong.'

Douglas followed Hoyle's advice for the initial phase of his career, but was destined never to venture into cosmology thereafter. Under the supervision of Roger Tayler, Douglas studied the properties of convection under the physical regimes found in astrophysical objects like stars.[1] In particular, he looked in detail at how convection was affected by, and reacted back upon, pulsation, magnetic fields and rotation. The theory he later developed as a postdoctoral research fellow to describe the interaction of pulsation with convection is still in use today.

In 1970, Hoyle suggested Douglas look at the impact of an effect called gravitational settling upon the solar neutrino problem. Settling is an effect we shall not dwell upon here: it will receive lengthy consideration in Chapter 8. This work got Douglas into thinking at length about the solar neutrino problem and therefore the interior structure of the Sun. Prior to Hoyle's prompting, Douglas had not computed a full solar model – the main thrust of his research had simply been elsewhere, in the study of classical pulsators like Cepheids.

Douglas began to ponder other ways of altering the modelled interior structure of the Sun to reduce the number of neutrinos produced. He soon hit upon a possible idea. This was based on his finding that the Sun might go through periods when its core would begin to oscillate.[2] The onset of these buoyancy-driven oscillations would mix the material in the core, acting like a 'solar spoon' (the phrase Douglas coined). This would reduce the rate of production of neutrinos and also lead to a lowering of the solar luminosity. Furthermore, the calculations suggested it

[1] The story is told that Tayler and Hoyle flipped a coin to decide who would supervise Douglas and who would take on their other new PhD student, Bob Stobie.

[2] The origin of these oscillations shared much in common with the 'epsilon' mechanism Eddington had proposed as one possible cause of pulsations in Cepheid variables.

would be an intermittent effect, cycling on and off roughly every 200 million years. This timescale was of some interest, for it corresponded roughly to the separation in time between extended periods of suppressed temperature – ice ages – on Earth. It was tempting to suppose there might be a causal link *if* the phenomenon was indeed acting.

Douglas did the initial work with his PhD student, Fisher Dilke. But it was during the next phase of the investigation, conducted with new student Jørgen Christensen-Dalsgaard, that things really got interesting.

Jørgen had studied astronomy as an undergraduate at the University of Aarhus in Denmark. There, he had developed an interest in complex material motions found in astrophysical bodies, convection in stars being a prime example. However, there were no clear opportunities in Denmark at the time to pursue graduate studies in theoretical astrophysics, and so he began to look abroad for suitable openings. An obvious location was Cambridge. Jørgen's family took the short trip across the North Sea each autumn for a regular break, and they had friends studying in Cambridge too.

In 1972, Jørgen began what he calls a 'fairly unsystematic' search for research opportunities in Cambridge. His first port of call was the Department of Applied Mathematics and Theoretical Physics. There, he was able to arrange a short interview with the then Head of Department, George Bachelor. As Jørgen recalls, 'probably not surprisingly he was not very welcoming to this Danish student' who had appeared unannounced 'off the street'. Jørgen was encouraged to take his application further through the official channels.

Shortly after this interview, Jørgen had the chance to meet a graduate astronomy student named Ian McIvor. McIvor suggested he speak with a young lecturer at the Institute of Astronomy called Douglas Gough, who held a joint appointment at the Department of Applied Mathematics and Theoretical Physics. Gough was pursuing research that overlapped nicely with Jørgen's interests. Jørgen managed to track down Douglas in the nick of time, on the very final day of his stay that year in Cambridge. In fact, he ended up discussing the possibility of joining Douglas on the way to the ferry in Harwich. During their long chat Douglas was able to form enough of an impression of Jørgen to realise this was a chap he really had to take on. So it was that in the autumn of 1973 Jørgen joined the Department of Applied Mathematics and Theoretical Physics to work with Douglas.

After finding a sign error in a term in the code Fisher Dilke had written to model the solar spoon – which meant the results then made a good deal more sense – Jørgen set to work on a new, improved and more realistic calculation that would take into account energy exchange between the proposed core oscillations and the environment in which they were supported. The first results of this 'non-adiabatic' analysis were presented at a conference in the summer of 1974. It was at this point Douglas and Jørgen realised they would require better solar models to take the study further.

As things stood they were still not in a position to be able to say whether the solar spoon really was able to stir the core as proposed. It was now clear that insufficient precision in several aspects of the then best solar models was becoming a real hindrance to reaching any firm conclusions. They therefore began a project to write a new, more precise stellar evolution code, which included an improved treatment of the physics involved. The aim was still to use this to test the power of the spoon. But that changed in the summer of 1975.

In June of that year Douglas hosted a conference on the theory of stellar convection at the Institute of Astronomy in Cambridge. A break in the conference was to provide a dramatic interlude as Henry Hill paid a visit to the Institute to present his latest results. Hill, who was working at the University of Arizona with Tuck Stebbins and Tim Brown, claimed to have discovered oscillations of the whole Sun in measurements of the apparent solar diameter. The ten or so modes of pulsation Hill said had been uncovered possessed periods ranging from approximately 7 all the way up to 50 minutes.

This naturally triggered the extreme interest of Douglas and Jørgen. Because of the pulsation studies they had been conducting it was possible for them to calculate a set of frequencies to compare with Hill's observational data. It was a small matter to modify the code to output frequencies of acoustic, whole-Sun pulsations in addition to those of the buoyancy-type oscillations driving the solar spoon. Jørgen spent that evening computing the required numbers. In a talk given the next day he was able to demonstrate that agreement between the claimed oscillation periods and those of the model was extremely good. The modelled pulsation frequencies that matched best were the overtones of some of the low-degree $l = 0$, 1 and 2 acoustic modes.

This story was to have a disappointing outcome in one regard, but extremely beneficial and far-reaching consequences in another. It is now generally accepted that the oscillations claimed by Henry Hill were instead signatures of atmospheric noise. Jørgen also notes that the solar models, although cutting-edge at the time, were rather crude compared with those in use today. In hindsight, the excellent agreement of the predictions and observations can therefore be seen as an unfortunate coincidence. In spite of this, Hill's presentation inspired Douglas and Jørgen and focused their attention on a wider problem. They now set to thinking through the potential science benefits of having a set of real solar mode data like these available.

Previous observations of globally pulsating stars had revealed objects that were oscillating in one mode or at most a few. The possibility that a single star might oscillate in ten or more modes simultaneously was a different matter entirely. The potential information content of the observations would then be so great as to allow various aspects of the interior structure of the star to be re-constructed. Furthermore, any inferences drawn would be far less sensitive to shortcomings in the detailed assumptions used to construct the solar models.

Excited as they were by the prospect of being able to make detailed seismic studies of the Sun – in essence by adopting an analysis philosophy similar to that in geoseismology – Jørgen and Douglas wrote a paper for *Nature* called 'Towards a Heliological Inverse Problem'.[3] They drew on the observations of Hill and colleagues and the 160-minute observations of the Crimean and Birmingham groups and discussed the possible use of 'inverse' techniques to infer the solar structure. But what is an 'inverse' problem? And, for that matter, is there an alternative 'forward' problem?

Let us begin with the second question. The answer to this is yes. In fact, the forward problem is the one I believe most people would think of first tackling. A classic example would be the following. Compute the resonant frequencies of a solar model. Then compare these with a set of frequencies extracted from observations. Now tweak the input parameters of the model but in such a way that key gross solar characteristics, such as the luminosity and radius of the star, match the known values – the model must give the observed gross characteristics if it is to be a plausible one. Then re-compute the frequencies and compare again. In this way it is possible to explore a grid, or parameter space, of model input parameters and their changing relation to the observations.

As the name suggests, the inverse problem looks at things from the opposite direction. The observed frequencies are a physical manifestation, or output, of the conditions found within the solar interior. They are determined by the integral properties in the volumes that the waves forming the modes occupy. An inversion seeks to re-construct a model of the interior – for example, the trend of sound speed – that will give rise to, and therefore be consistent with, *all* the observed frequencies simultaneously. The key to this approach is having data for a selection of modes that have different cavity sizes in the interior and therefore penetrate to different depths. One is, however, limited to a re-construction of those parts that are sampled by the observations. The more plentiful the modes, and the more varied their depths of penetration, the better.

To see how this works, let us start with just two modes. We select modes that have very similar properties from the surface down through

---

[3] J. Christensen-Dalsgaard and D.O. Gough, *Nature*, 259, 1976, p. 89.

their interior cavities. Any change to the physical characteristics somewhere in this common layer will give a similar change in, say, the frequency of each mode. That said, we make sure there is one crucial difference: the first mode penetrates a little more deeply than the second. If we measure the difference in frequency we will therefore get some idea of the physical conditions in the thin layer the first mode samples, but the second mode does not.

In practice the use of just two modes would normally give only a very crude measure. This is because the difference in frequency would unfortunately depend on more than just the conditions in the deep layer seen by the first mode. There would be contributions from elsewhere in the cavities too. The difference would therefore represent a potentially confusing measure of properties from a selection of depths.

To get around this problem one requires many modes, to sample as diverse a range of cavities as possible. By making combinations of several, not just two, modes it is possible to get a measure that really does depend on conditions only in a particular layer of the interior. Then, by varying the mode combinations, different layers can be probed and a picture of the interior structure as a function of radius may be formed.

An inverse analysis of this type requires we know how the various modes sample the interior structure. This is not simply a matter of cavity size. The detailed properties of each mode are important – for example, where the nodes and anti-nodes are located. Changes located right at a node will leave the mode in question unaffected, and unusable as a probe of conditions there. The detailed response functions of the modes, once calculated, are called kernels. These can be produced for a variety of different observed parameters.

Inversion analyses of this nature are used in many areas of science – for example, to interpret the interior structure of the body from a computed tomography (CT) scan,[4] or to infer the properties of the Earth's interior from seismic data. However, inverse problems often suffer from

---

[4] A CT scan uses X-rays to probe the interior structure of the body.

a fundamental underlying problem. A given set of observational data does not necessarily lead back to only one solution. This is not an issue of quantity or quality – even with an infinite amount of data collected with infinite precision (unrealistic for sure), it may be that more than one solution, or inferred model, will satisfy the data. In mathematical parlance the solution is said to be 'non-unique'.

This famous problem was posed in 1966 by the mathematician Mark Kac in this way: 'Can you hear the shape of a drum?' Mathematicians Carolyn Gordon, David Webb and Scott Wolpert solved this tricky question in 1991. They found two 'mathematical drums' with distinctly different shapes that had the same set of resonant frequencies. The solution was not unique. At first glance this all seems rather worrying. How much store can we place in inversions of helioseismic data? In answer, it must be recognised there are differences between the problem posed by Kac and the solar problem of interest to us here. Even if our solar problem were simplified, for example by making the cavity boundaries perfect reflectors (as in Kac's problem), or by having infinite data available, it would remain unclear whether or not there is a unique solution. We simply need to be sensible; to remember there might be a problem (then again, there might not) and take the interior structures we recover with a tiny pinch of salt.

The possibility of making inversions of data was outlined in the 'heliological' submission to *Nature*. The editor was at once interested. In its initial form the paper contained no reference to the 160-minute oscillations. However, the journal drew these observations to the author's attention, and the resulting implications were then added to the paper before its appearance.

At this time, the 160-minute signal was still regarded as solar in origin. Certainly the significance of the pulsation found in the Crimean, Birmingham and Stanford data was beyond doubt. Andrei Severny and George Isaak had both speculated, independently, that the uncovered signal was in fact the fundamental radial acoustic mode of oscillation.

George had been sceptical that it could be a gravity mode. He knew one would expect the spectrum of gravity modes to be very dense, so that many modes would be packed into a narrow frequency band of the spectrum. His objection was, therefore, that some mechanism would be needed to preferentially excite one mode at the expense of all the others to explain what was seen.

In opting for the breathing-mode explanation, the Crimean and Birmingham teams were both aware of the consequences this would have for the structure of the deep solar interior. We know already that a rough estimate of the period of the fundamental acoustic mode can be made from the sound travel time across the star. Temperatures higher than expected in the interior would have the effect of raising the sound speed. This would reduce the sound travel time and the fundamental period. On the other hand, lower than expected temperatures would reduce the sound speed and therefore increase the period. Since the observed 160-minute signal had a period rather longer than the expected value for the fundamental, of about an hour, one conclusion was that the core temperature was lower than had previously been thought – enter once more the solar neutrino problem. However, as Douglas pointed out, this would also have consequences for one of the fundamental observational constraints any model must match – the solar luminosity. The lower core temperature needed to explain the 160-minute period would reduce the luminosity significantly below the value that is observed. This interpretation therefore seemed to be the wrong one.

With the radial-mode explanation ruled out, the question of the origin of the 160-minute signal remained outstanding. In the years that followed, improved analyses of the data suggested the signal was not, after all, solar. One of the concerns since the first detection had been how close the period was to the ninth harmonic of the day (24 hours divided by 9). Re-analysis suggested the measured period was in fact precisely that of the ninth harmonic, not ever so slightly higher as had first been thought. It was then possible to show that the observed signals could be generated as an artefact of a day–night modulation of the observations,

Figure 6.1. Trying not to get too close to the lava! On Mount Etna, during a break in a meeting, 1982. Left to right: Douglas Gough, Franz Deubner and Wojtek Dziembowski. Picture courtesy of D.O. Gough.

as given by data from a single ground-based location. Today this is still regarded as the most likely explanation for what was observed.

It is ironic that the two observational results that acted as the inspirational springboard for helioseismology were, most likely, not solar oscillations after all – one probably noise, the other a subtle observational artefact which took over ten years to pin down. But the field was built on firm foundations. The true nature of the 5-minute signal was uncovered in 1975, and confirmation of the whole-Sun character of some of the modes followed in 1979. It was time to start getting science from the 5-minute modes.

The first comparison between the observed and theoretically predicted locations of ridges in the k–ω diagram had been made by Roger Ulrich. In the mid 1970s, Hiroyasu Ando and Yoji Osaki made an improved calculation determining frequencies for an envelope model of the Sun.[5] They found that if they used then accepted physical parameters to build their model of the Sun they could not match the observed positions of the ridges.

Douglas Gough learned of Deubner's data too late to incorporate them into the 'heliological' paper. He was, though, intrigued by the mismatch between these data and the predictions of Ando and Osaki. The urgency with which he set to work to find a possible solution was motivated, at least in part, by an invitation to speak at a conference the date of which was rapidly approaching. The basic technique he used in this study has been refined to a high level of sophistication since.

Douglas was interested in looking at the effect of small changes, or perturbations, to the model of Ando and Osaki. One way of doing this would have been to first repeat, in its entirety, the analysis of the Tokyo-based scientists. Though this was the thorough thing to do, it would certainly have taken far too long. Instead, Douglas realised he could take a short cut that would not significantly compromise the robustness of any conclusions he might draw. This involved using a simple model of the outer layers of the Sun – a polytrope (Chapter 2) – to make predictions of the effects of perturbations in the structure.

Ando and Osaki had calculated the resonant properties of a real solar model. What Douglas did was to approximate the surface of the Sun with a polytrope and then ask the question: by how much do I need to change that polytrope to bring Ando and Osaki's results roughly into line with the observations? He was thereby using a simple model to calculate the effects of small changes to the full calculation.

Douglas had just written a theoretical paper with Nigel Weiss (also at Cambridge) about the depth of the solar convection zone. This had presented a relation between the properties of the polytrope and the depth of the zone. So, having determined the extent to which the polytrope had to be altered to match the observed ridges in the k–ω diagram, it was then

[5] H. Ando and Y. Osaki, *Publications of the Astronomical Society of Japan*, 27, 1975, p. 581.

a straightforward matter to use the result from the paper with Weiss to infer a depth for the convection zone. The result was that the depth needed to be about 225,000 kilometres, approximately one-third the radius of the Sun. This depth was about fifty per cent larger than typical values used in solar models of the time. (Modern analysis fixes the value closer to 200,000 kilometres.)

Ed Rhodes and Roger Ulrich had realised, independently, that the first important result they could get from their data was – the depth of the convection zone.[6] They compared the observed frequencies with predictions from their own models of the solar envelope, and found, like Douglas, that the results implied a deep convection zone. It is worth adding that in his groundbreaking paper in 1970 Roger had stated, 'we may hope that improved observations of the 5-minute oscillations will allow an accurate determination of the envelope' conditions.[7]

The result was a first – an estimate of an important internal structural characteristic of the Sun constrained directly by helioseismology. The prediction would have immediate implications. First (and unbeknown to Douglas at the time) it solved some of the problems of Peter Gilman. Gilman had been working on models of convection in the rotating Sun. If the depth of the zone was fixed at a shallower value he found he was unable to get anything like the observed pattern of surface rotation at high latitudes. With a significantly deeper zone the results were more encouraging.[8]

Second, it appeared to make it harder to find an astrophysical solution to the stellar neutrino problem. The stellar modellers had been trying to push the solar model parameters as far as was deemed reasonable in a direction that would have reduced the accepted value for the central temperature of the Sun. This would have resolved the solar neutrino problem, because a lower core temperature would have reduced the flux of neutrinos. Lower internal temperatures would have implied,

[6] R.K. Ulrich and E.J. Rhodes, Jr, *Astrophysical Journal*, 218, 1977, p. 521.
[7] R.K. Ulrich, *Astrophysical Journal*, 162, 1970, p. 993.
[8] See Chapter 10 for a discussion of rotation in the outer layers.

among other things, a lower abundance of helium throughout the star – which in turn gave a shallower convection zone.

The prediction from helioseismology went in the opposite direction. It suggested the convection zone was rather deep. The natural conclusion was that this made the solar neutrino problem seem much worse from a solar modeller's perspective, but with an added caveat. The inference made was about the helium abundance in the outer parts of the Sun, not deeper down in the energy-generating core. The two were unlikely to be very dissimilar, but what was needed were data on the modes penetrating the core itself.

Subsequently, Douglas spent almost a year on sabbatical in Nice, working with Yoji Osaki, Janine Provost, Gabriel Berthomieu, Arlette Rocca and his student, Alan Cooper. Rather than use the simple polytrope, the aim was to do a proper calibration to confirm the new estimate of the convection zone depth. After the sensitivity of the result to a large number of different factors had been tested, the 'deep' estimate was upheld as robust.

Following from this first major success for the field the main thrust of Jørgen and Douglas's work became to obtain the best-calibrated solar model possible. They began to look in detail, with Guy Morgan, at models with different abundances of helium and heavy elements – the two are closely linked in stellar evolution theory. An important goal of this work was to assess the impact of these different interior compositions on the number of neutrinos emitted by the nuclear fusion reactions in the core. In particular, Jørgen and Douglas were looking to see how the oscillation frequencies of such models might vary with the composition – might they change by enough to allow the different models to be discriminated against? The paper that resulted from the initial phase of this work, submitted in the summer of 1978, was therefore the first to address in detail whether helioseismology could shed light on the solar neutrino problem.

An important motivation for their investigation was earlier work by Icko Iben, and also John Bahcall and Roger Ulrich, which showed that the helium and heavy element abundance put into models of Sun-like stars could have a notable impact upon the flux of neutrinos generated. To produce a good solar model everyone was aware that the choice of abundances had to be made subject to a very important constraint.

We have seen that although one cannot measure directly the abundance of helium in the solar atmosphere it is possible to get an estimate of the percentage abundance of heavy elements. In the late 1970s the accepted value for this was about two per cent. This is one of the main input constraints for the models. But this one number might not tell you all you need to know.

Observation gives a measure of the quantity of metals in the Sun's atmosphere. But what if this fails to give a reliable estimate of the amount deeper down? Consider the following scenario. As time ticks by we know the Sun loses a small amount of mass via the solar wind. However, it could also be *taking on* a little bit of mass from the surrounding interstellar medium. This is the stuff between the stars – the vast expanse of gas and dust from which the stars are born and into which they cast their material when they die. Since this matter contains the leftover remnants of dead stars, and the material they have processed by nuclear fusion, metals will be present. If some of this material were to be swept up – or accreted – by the Sun, its atmosphere would become enriched in metals. If this material could not be efficiently mixed from the photosphere down into the deep interior, the atmospheric abundance might mislead an observer into thinking the quantity of metals in the deep interior was equally as rich.[9]

The concept of heavy element accretion had been around for many years. Fred Hoyle had discussed it in the late 1930s. However, the advent of the solar neutrino problem – with the work of Ray Davis, John Bahcall and others in the 1960s – had reinvigorated interest, for this concept provided a possible way to solve the solar neutrino problem. If the deep interior of the Sun had a lower metallic content than previously

---

[9] Any material accreted will be well mixed in the convection zone.

thought – again, because the observed surface value had been augmented from outside by accretion – the opacity of these layers would also need to be revised downwards. Remember that opacity measures how hard it is for energy to flow by radiation. We have seen that metals give high levels of opacity because they offer a multitude of ways in which packets of radiation (the photons) can interact and so be removed from the outward flow of radiation. A widespread reduction in opacity, over a range of temperatures encompassing the hydrogen-burning, neutrino-generating core, would allow the radiation to flow outwards more easily. A reduced core temperature would then suffice to give the luminosity observed at the surface. And lower core temperatures would produce fewer neutrinos.

In order to understand whether this was a viable solution to the solar neutrino problem, Douglas, Jørgen and Guy looked at three solar models. Their 'sequence' included a standard model, with an initial heavy element abundance of two per cent throughout the interior; and two further models – one having an extremely low initial abundance of 0.1 per cent, the other of 0.4 per cent – which they allowed to accrete matter over the simulated aeons so that by the time they reached the current age of the Sun the abundance in their atmospheres matched the observed value of two per cent.

It was found that a model with an initial low abundance did indeed give reduced neutrino rates roughly in line with the measurements of Ray Davis's chlorine detector. This was to be expected. However, the key part of the work was then testing the effect upon the resonant properties of the models. The scientists found that the frequencies of the oscillation modes differed sufficiently between the three models to offer in principle a means of picking the best one. However, data on the right modes were needed to do the job.

The only data available when this work was done were those for the high-degree, high-wavenumber modes that penetrated only a few thousand kilometres below the surface. We have seen already that these favoured a deep convection zone. This meant they were also at odds with a low surface abundance of helium – a deeper zone goes hand in hand with higher helium content. Given that the amounts of helium and heavier elements are tied together, this seemed to suggest the heavy element abundance could not be low as well.

Although these shallow-penetrating modes were therefore sensitive to the abundance, assessing the fine detail was fraught with difficulties, since any model calculations of the frequencies had to rely on how the top of the convection zone was modelled. This is the part of the interior where the sound waves begin to interact much more strongly with the solar plasma. By this we mean the exchange of energy by radiation starts to become important and cannot be ignored. A proper treatment then demands that a non-adiabatic description be employed. Deeper down this energy exchange can be safely ignored, a simpler adiabatic description suffices and as a result modelling the oscillations is trivial by comparison. The problems inherent in modelling the near-surface layers – the outer few hundred kilometres of the convection zone and above – are still with us today.

Modes that would penetrate the deep interior and core were, however, another matter. The uncertainties inherent in modelling the very near-surface layers contributed a smaller fractional error to the calculated frequencies of these low-degree modes. Furthermore, from a common-sense perspective it was these modes that would provide a *direct* probe of the deep interior, the part of the Sun which bore the scars of evolution most conspicuously, and the part whose properties would be most affected by changes to the abundance.

In 1978, the 160-minute oscillation was still prominent in the Crimean, Birmingham and Stanford datasets. Henry Hill also continued to report evidence for whole-Sun oscillations in his solar diameter data. However, even if these various detections were regarded as possibly being of real solar modes, the identification of exactly what they were remained very uncertain. To their credit, Douglas and his colleagues urged considerable caution in using these apparent long-period modes to address the abundance question. But any reluctance to use observational data was swept away when, in 1979, the whole-Sun nature of some of the 5-minute modes was uncovered by the Birmingham group.

In order to use these core-penetrating data to their full potential the observed modes had to be identified. Without knowledge of the angular degree, l (which would give the equivalent horizontal wavenumber), and overtone number (the radial order, n) of each observed oscillation peak it would not be possible to describe in detail which parts of the interior the modes probed. Without that information a precise picture of the interior would not be forthcoming. The identification was in practice done in two stages. The first part – tagging the angular degree of each mode – was much easier and so happened rather more quickly. However, this was only possible after the South Pole dataset of Eric Fossat and colleagues had been collected. The power spectrum of their observations was dominated by the familiar picket-fence pattern, but with the intrinsic high resolution this dataset provided it was now possible to see that each plank of the fence was actually made up of the expected pairs of modes – $l = 0$ and $l = 2$ on the one hand (the even pairs) and $l = 1$ and $l = 3$ on the other (the odd pairs).

Eric divided up the spectrum into chunks so that each length contained the same types of peaks. He then overlaid the pieces – a 'superposed frequency analysis' – so that the pattern built up in strength against the noise. It was then possible to discriminate the different types of modes present in the spectrum. Their angular degrees could be identified from the spacings between and the relative strengths of the peaks in the overlaid data. This required some theoretical input. Predictions of the frequency spacings were available from detailed model calculations. They now also came from the approximate expressions given by asymptotic analyses,[10] which worked well for the observed *high* overtones (as we mentioned they do, in Chapter 3, for modes of a cone or uniform sphere). Asymptotic analysis for helioseismology has been further

---

[10] Monique Tassoul made an in-depth study of the application of asymptotic analyses for non-radial oscillations of stars: M. Tassoul, *Astrophysical Journal Supplement Series*, 43, 1980, p. 469.

developed since by the likes of Ian Roxburgh and Sergei Vorontsov. Another important contribution came from detailed mode visibility calculations performed by Douglas and Jørgen.

With this analysis complete, it was now possible to say whether a peak in the South Pole or Birmingham data was the signature of an $l = 0$, 1, 2 or 3 mode. Robust identification of several $l = 4$ and $l = 5$ modes came shortly after, in 1981, when Phil Scherrer analysed the latest Stanford data. Phil had plots of the data with him when he travelled to a meeting in the Crimea in September of that year. The reader will recall that the Stanford instrument, like its Crimean counterpart, measured the difference between the Doppler shift of a central patch and a region around the rim of the solar disc. Because the central part covered only about twenty-five per cent of the area of the disc the observations were sensitive to modes with slightly smaller horizontal wavelengths – i.e. slightly higher angular degrees – than the Sun-as-a-star data.

At this meeting Eric presented the results of his superposed frequency analysis. He also presented an echelle diagram of the data. This is made from the same pieces, or strips, of spectrum used to perform the superposed analysis. Rather than being added together, the strips are arranged vertically (Figure 6.2).

Imagine doing this on a piece of paper. The first strip lies at the bottom of the page. The strip that comes next in frequency is positioned above the first strip, so its start and end points (in frequency) align vertically with those of the first piece. The next strip goes above this, and so on. If the spacing of peaks in the spectrum is the same throughout, vertical lines of peaks will be conspicuous on the page, each line corresponding to a different angular degree. In reality, subtle changes in spacing in the solar spectrum cause the lines to bend – however, the structure is very apparent and provides a simple, but efficient, way of identifying the degree of each mode. The human eye and brain have a wonderful capacity for recognising patterns of this type, particularly if the data are quite noisy.

Impressed by Eric's results, Phil managed to get hold of some large Soviet graph paper and, with his Stanford spectrum to hand, set to work

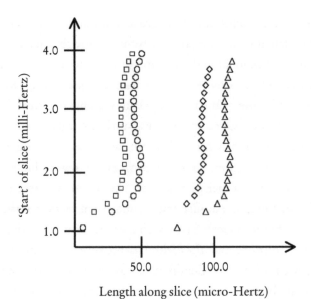

Figure 6.2. An echelle diagram, showing the location of low-degree, core-penetrating acoustic modes in frequency space. The frequency spectrum is split into slices of a suitable length; each then occupies a horizontal strip on the diagram. Rather than showing a diagram made from the raw spectrum, we have used instead symbols to show the locations of the mode peaks. Locations of $l = 0$ modes are plotted as circles, and $l = 2$ as squares; $l = 1$ modes are plotted as triangles and $l = 3$ modes as diamonds. The data come from modern observations, which can detect modes all the way down to a frequency of about a milli-Hertz.

over a lunchtime constructing his own echelle plot. He did so by marking the locations of the peaks on the paper (as in Figure 6.2). When the plot was complete he showed it to Douglas, who observed, 'That line is $l = 3$, that one 4 and the other 5.' The degrees of an extra cohort of modes had been tagged.

Douglas and Jørgen were able to get a rough idea of which of the models fitted the whole-Sun data best from a quick comparison of the observed large mode spacing in the Birmingham data – the separation in frequency between consecutive odd- or even-degree mode pairs – with that of their own model calculations. A much more refined, though still fairly rough, comparison was then made with estimates of the actual mode frequencies, published independently by the Birmingham and Nice groups in 1980.

What made a proper comparison difficult was that the overtone numbers of the modes were still awaiting unambiguous identification. Because of this uncertainty the analysis threw up two plausible solutions – one with the low helium abundance, the other with a high value of twenty-five per cent. Douglas and Jørgen concluded the latter was most likely the better match, because they could bring to bear conclusions made from the higher-degree data which pointed strongly to there being a deep convection zone and high helium content in the outer layers.

So, the whole-Sun, low-degree data had made an important early contribution, which suggested low heavy-element models could not be invoked to solve the solar neutrino problem. Once the modes had been properly identified, it became possible to make the case with even more conviction. Standard models seemed to provide the best match. But this second, trickier stage of the identification process only followed thanks to observations made by Jack Harvey and Tom Duvall, which bridged the gap between the low- and high-degree modes.

Tom Duvall took his PhD at Stanford in the 1970s, where he worked under the supervision of John Wilcox, measuring material flows and magnetic fields across the solar surface. He shared an office with Phil Dittmer, who was involved with Phil Scherrer and John Wilcox in observing the 160-minute oscillation with the Stanford solar telescope. This meant Tom was fully aware of the advances being made in the fledgling field of helioseismology, and he too became actively involved after taking up a postdoctoral position at Kitt Peak. At first Tom continued to study surface flows on the Sun. His focus shifted after he crossed paths with Jack Harvey and Ed Rhodes.

Jack – who we recall had already more than dabbled in helioseismology through his involvement with John Leibacher in HOWL – continued to keep tabs on the field, including a regular correspondence with Franz Deubner. Ed, who made his breakthrough observations at Sacramento Peak, had wanted to find a way to continue collecting data. This proved a key stimulus for Jack. He had an early CCD (charge coupled device) camera, which could be used to image light from the Sun onto many resolution elements, called pixels. This CCD (then called a CID or charge injection device) could therefore form the heart of an instrument to look in detail at the oscillations.

And so, with a feeling that 'we could really contribute to observing these new-fangled oscillations', Jack set to work. With the onus on him to develop the instrumentation, he was going to be hard pushed to find time to analyse the data fully as well. It was at this point, after chatting things over with Jack and Ed, that Tom offered his services. He would do the analysis.

Jack and Tom started collecting data with the McMath–Pierce Telescope, at the Kitt Peak Observatory, in the summer of 1978. They began by duplicating the high-wavenumber, high-degree observations of Deubner and Rhodes et al. looking at modes of l greater than about four hundred. In early 1980 they upgraded the CID camera and began to collect data on modes with l as low as 150. Further observations, made with Ed Rhodes, concentrated on looking for changes in the rotation in the near-surface layers.

Ed was also in the early stages of developing an observational programme. He had started teaching at the University of Southern California in late 1978. The promise of a tenure-track position a year later gave him the leverage needed to start his own observations. Soon he got agreement from Bob Howard to begin work on setting up a permanent capability to observe high-degree modes from the Mount Wilson Observatory (where Bob Leighton and Bob Noyes had made the original observations of the oscillations).

Meanwhile, by late 1980, Jack and Tom had realised that looking at the entire disc of the Sun would be extremely valuable. They decided to do so at a high resolution and not in a Sun-as-a-star manner. By imaging the visible disc onto many pixels they would be able to observe modes covering a vast range in angular degree. But with a large number of pixels covering the disc would come a huge computational burden to calibrate the raw data. Jack and Tom knew they would eventually have to run such a system in order to extract the full potential of the oscillations. But in the meantime Jack made use of an ingenious scheme that reduced substantially the task of analysing the data. A cylindrical lens introduced at the front of the instrument collapsed the image of the Sun onto a 'line' (or narrow bar) and meant the data were sensitive only to a particular subset of modes.[11] The orientation of the lens with respect to the Sun's rotation axis determined which modes were left in the data. Positioning the lens either parallel or at right angles to the rotation axis gave two respective datasets with rich scientific potential.

One orientation left modes in the data ideal for probing the trend of sound speed through the solar interior, these being mode components that suffered no frequency shift from the rotation. The other set was perfect for re-constructing the internal rotation rate in successively deeper layers in the solar equatorial regions; it was made up of the outer mode components that suffered the largest frequency shift from the rotation.

The data for both studies could in principle be extracted quickly, since the numbers of modes needing to be analysed had been reduced, relative to the complete full-disc set, by well over two orders of magnitude. (The large number of missing mode components that would have been present in the full-disc data were the other rotationally split components, with their intermediate-sized frequency shifts.)

The new system was run for the first time in June 1981. During the remainder of the summer and the autumn Jack and Tom made observations of medium- and low-degree modes. The high-degree collaboration with Ed also continued. In 1982, attempts were made to collect an extended, multi-day set of observations. Alas, an unlucky run of poor

---

[11] In the earlier observing runs, a glass block was used that in effect served the same purpose of selecting out certain modes.

weather intervened. Another attempt, in May 1983, bore fruit – a crucial, 17-day sequence of high-quality data.

The system required an observer to be present at all times – to be up nice and early to get the equipment properly configured, and to monitor its operation during the course of the day. Anyone who has been involved in observing activity of this type, when a single astronomical object is monitored for long periods of time, knows that success comes at a cost. For 'success', read: 'things are running smoothly'. The cost is that there is then very little for the observer to do. Solar observations have the advantage that one does not have to battle with a disrupted sleep pattern. However, Jack now recalls with some amusement – and he can do this because they are the best of friends – that he and Tom began to get on each other's nerves as day after day of lovely, clear sunshine came and went. Eventually it was agreed they would do observing duty on alternate days.

The data they had collected since 1981 were 'a bridge in a gap in solar oscillations'. This is the phrase Douglas Gough used to describe the observations in a news and views article that accompanied the first important publication of Jack and Tom's work in *Nature*.[12] The description was apt. The Kitt Peak instrument was able to plug the gap because it could observe some of the low- and the high-degree modes already detected by others *and* many previously unseen modes in between. This meant a way had been found to link the observations from one extreme and another. But what proved crucial was the ability of Jack and Tom's instrument to observe a particular type of mode – the 'f' mode. With these data in the bag a complete identification of the overtone numbers of all modes could be made.

F modes are formed by waves that ripple across the surface of the Sun. The 'f' indicates they are *fundamental* gravity waves. They are like the waves observed on the surface of water and have no nodes in the radial direction.

---

[12] T.L. Duvall, Jr and J.W. Harvey, *Nature*, 302, 1983, p. 24.

When a stone is cast into a pond a surface ripple will spread out from the point of impact. This surface wave is a mixture of transverse and longitudinal disturbances. The behaviour of the wave depends upon how the wavelength on the surface compares with the depth of the water. If the wavelength is much greater than the depth the propagating disturbance is called a shallow-water wave. If, on the other hand, the wavelength is somewhat shorter disturbance is a deep-water wave and will penetrate to a depth of about one wavelength. In the latter case, the dispersion relation – describing how frequency varies with wavelength or wavenumber – then depends only on the local gravity and has nothing to do with the physical characteristics of the water.

This is an extremely useful property. If the deep-water condition could be applied to surface waves on the Sun, an estimate of the gravity at the surface would then allow the dispersion relation to be defined. This would then give an accurate prediction for the location of the f mode ridge in the k–ω diagram. If one could identify which curve this was, tagging the p modes with their appropriate overtone number (n) would then be a trivial exercise. The next ridge up in the diagram would have to contain single-overtone p modes, the one after that double-overtone modes and so on.

The waves cannot know anything about the Sun's curvature if the deep-water analogy is to hold. This is because the simple dispersion relation only works in the plane-parallel approximation. In this, the atmosphere within which the wave travels must look like a rectangular slab with no curved boundaries. Local gravity is assumed to act at right angles to the surface of the slab and to be constant across its depth. On the Sun these requirements are well approximated for waves with high wavenumber; because they have such short wavelengths over the surface the effects of the Sun's curvature can be disregarded. Jack's instrument could observe the f modes formed by these waves. He and Tom could therefore identify the overtone numbers in their data. A check of these results then allowed the low-degree data to be similarly tagged by others. The calibration problem had been solved.

There is a final anecdote to this story. Douglas Gough was fully aware of the significance of the deep-water approximation and had pointed this out in one of his papers. He was in the habit of using the resulting dispersion formula to check the calibration of the results of observers. When Jack and Tom published some of their results Douglas duly checked the data and was delighted to find they matched the prediction. When Douglas next spoke to Jack he indicated how impressed he had been that the observations had fitted the little-known formula. 'Oh yes,' said Jack, 'we found that formula in one of your papers.' He then told Douglas how they had used the formula to check the calibration of the instrument. An interesting case of observers picking up a remark in a theorist's paper and using it to cross-check their observations.

7

TAKING IN THE SUN'S RAYS

Jack Harvey and Tom Duvall had wasted little time after their initial success at Kitt Peak. Over the summer and autumn of 1981, in addition to
continuing observations at Kitt Peak with their newly installed cylindrical-
lens system, they had begun furious preparations for an observing run at
the South Pole (Figure 7.1). Jack had already worked with Eric Fossat,
who achieved spectacular success making observations of the whole-Sun
oscillations over the New Year of 1980. Come the austral summer of
1981/2 Jack, Tom and Martin Pomerantz were at the South Pole collecting data with an instrument that, having done away with the cylindrical
lens, imaged all parts of the visible disc of the Sun. This brought with it
the capability to observe every rotationally split component of a given
mode through a wide range in angular degree, something that had not
been possible when the cylindrical lens was in place.

   Jack and Tom's South Pole dataset was the first set of its type. These
'complete' data opened up the exciting possibility of re-constructing the
interior structure in latitude as well as depth, which would be extremely
beneficial for studies of the rotation below the surface. However, the
cylindrical-lens data were proving so scientifically rich their analysis was
still absorbing a large fraction of Tom and Jack's time. So it took a while

to get around to analysing fully their complete South Pole set. In the meantime other observers began to collect complete data.

Tim Brown's Fourier Tachometer instrument was a joint project between the High-Altitude Observatory in Boulder, Colorado, and the Sacramento Peak Observatory. The instrument – capable of observing modes up to angular degrees as high as one hundred – collected its first data in the summer of 1984. Tim's analysis of the Fourier Tachometer data provided the first look at how the interior rotation varied with latitude (of which more in Chapter 10).

Meanwhile, the extensive observations made by Ken Libbrecht over the latter half of the 1980s were to prove an invaluable resource for a rapidly expanding helioseismology community now hungry for data. Even though these observations were made from a single site – the Big Bear Observatory in California – they generated high-quality data that spanned a few months or so each summer. At the time, only the Birmingham and Tenerife groups were making regular, extended runs of this type – but for an entirely different set of modes, those with the very lowest degrees. Long datasets for the medium- and high-degree modes – degrees made accessible by the high-resolution detector in Ken's instrument – were not commonplace. In spite of the fact that his data could not probe the deepest parts of the Sun's interior, the sheer number of modes observed meant the data were ideal for probing conditions in the convection zone and the outer parts of the radiative zone. Furthermore, since Ken accumulated good-quality data over a period spanning several years, the data could be used to study in detail the effects of the 11-year solar cycle of activity on the properties of the modes.

Ken had honed his instrumental skills as a PhD student under the supervision of Bob Dicke. In the 1960s Dicke had begun an extensive series of observations that attempted to measure the shape of the Sun. These measurements will feature in Chapter 9. Come the early 1980s, Dicke wanted to take advantage of improvements in technology to repeat the observations. This work formed the topic of Ken's graduate thesis.

Figure 7.1. Tom Duvall (left) and Jack Harvey (right), and their instrument, at the South Pole in November 1981. Picture courtesy of J.W. Harvey.

These observations led Ken to become interested in helioseismology. To an outsider it appeared to be a vibrant and exciting area. A particular attraction was the youthful stage the field was still in – no one really knew what to expect from the next set of observations. After Ken had the chance to chat with Hal Zirin (from CalTech) about his ideas on making observations, Zirin promptly offered Ken a job and access to CalTech's solar observatory at Big Bear.

The strategy Ken opted for was to make imaged, resolved observations of the Sun. Even though he was aware Jack Harvey, Tom Duvall and Tim Brown were already in the game, he felt it too exciting an opportunity to pass up. His basic working philosophy would be to 'look at as many modes as possible for as long as possible'. Construction of a dedicated instrument began in 1984.

An important part of the instrument was a special filter that picked out a narrow range in wavelength around a Fraunhofer absorption line formed by calcium. Ken intended to observe the oscillations by tracking the Doppler shift of this line. Zeiss Optics made the ideal filter for the job – but these filters came at what for Ken was the potentially prohibitive cost of about US$100,000 per unit. However, Hal Zirin had handily stockpiled a few spare ones and now loaned one indefinitely to Ken. With this key component in place, the instrument made its first extended observations in the summer of 1985.

A full summer run required the collection of about twenty thousand images of the Sun, each comprising a pixel-by-pixel measure of the Doppler shift across the visible disc. All modes were accessible at each angular degree – this gave access to a grand total of about ten thousand different components. Suffice to say, this represented an awful lot of information to extract from the data.

One of the most time-consuming elements involved transposing (re-ordering the elements of) a matrix of twenty thousand by ten thousand elements. Although this could be done by a brute force approach, it required a very large amount of computer disk space. Ken estimated he needed about a gigabyte. This may not seem like a lot now, in an age when off-the-shelf personal computers have a storage capacity of several tens of gigabytes. However, in the mid 1980s this was asking too much of the average computer. Ken's solution was to apply for time on a super-computer in San Diego.

Several of these machines had been set up to allow the scientific community to run calculations that demanded a number-crunching capability beyond what could be achieved with more widely available computers. Although Ken did not require a supercomputer's fast processing and large memory, he did require its large storage space.

Once he had been granted time on the machine, the next problem was how to transfer the data from his observatory to the supercomputer in San Diego. One option was to do this over the youthful Internet. The other was to put the data tapes in a car and drive down from Pasadena to

San Diego. In terms of speed of transfer, it really was a toss-up between the two. As Ken puts it, 'the freeway option had a good data-rate behind it'. However, the Internet transfer was marginally faster on paper – although it did mean Ken had to spend several days loading and unloading tape after tape of data.

The helioseismology community used Ken's data widely. In addition to illuminating the effects of the solar cycle and the rotation of the interior, the data had a pivotal role to play in fingering excitation by turbulence, at the top of the convection zone, as the driving mechanism for the modes. Come the early 1990s Ken decided to leave the field. By then he had collected data spanning roughly one half of an 11-year activity cycle (from the minimum level of activity in the mid 1980s up to the maximum at the turn of the following decade). He felt it might be tough to extract significantly more from these single-site data, with their long gaps in coverage each day, without continuing the observations for at least the same period of time again.

The community had by then reached the conclusion that dedicated year-on-year observations which were as near continuous as possible were a must. And so the field entered a new observational era. Before we return, in Chapter 8, to our historical account of the science landmarks of helioseismology, this chapter tells the story of how the major, permanent observational programmes were established.

The first serious discussions regarding the need for continuous observations originated in the early 1980s. Longer datasets meant it would be possible to measure the characteristics of the modes to a much higher precision. This would not only allow tighter constraints to be placed on the interior structure of the Sun but also raise the possibility of hitherto unseen features being uncovered, and new science following (which certainly turned out to be the case). A second potential benefit would be improvements in the accuracy of the measured characteristics. Accuracy is subtly different from precision. Precision bears on the uncertainty with which we can measure a quantity – for example, the frequency of one of the modes. Accuracy is all about how well our measurement matches the actual value of the parameter. When observations and the subsequent analyses needed to extract the parameters are complicated, good agreement cannot always be taken for granted. Any mismatch can lead to erroneous scientific conclusions being drawn from data.

Most of the data analysis in helioseismology is done on the frequency spectrum of the observations. The peaks of the modes dominate these spectra. The bread-and-butter analysis of the helioseismologist involves determining, very carefully, the properties of the peaks, the most obvious example being their frequencies. This is called 'peak bagging'. Values extracted by the analysis are then used as input to infer, say, the trend of sound speed in the interior.

The oscillation spectra have an extremely busy appearance, containing very many p mode peaks. Some of these lie very close together in frequency. Indeed, because the peaks have some width – the sound waves are damped – the peaks often overlap. This can make disentangling one from another, and extracting a reliable estimate of frequency, rather tricky. But lengthier observations bring with them an improved frequency resolution and this can help to alleviate the problems in that it becomes easier to obtain a frequency estimate less tainted by the unwanted influence of other nearby peaks. This is particularly important for attempts to measure the separations of nearby, rotationally split components.

In the early years the accuracy problem of most concern to helioseismologists arose from the presence of gaps in their observations. Data collected from a single terrestrial site contain large, regular gaps – the Sun has an annoying habit of setting on the observer. Over the course of a year there will be a seasonal variation of the time-span over which data can be collected each day, and breaks arising from inclement weather. However, the dominant pattern will be the daily on–off signature with its repeat period of 24 hours.

The p modes give strong peaks in the frequency spectrum because they are periodic. Alas, the mathematical routines used to compute the

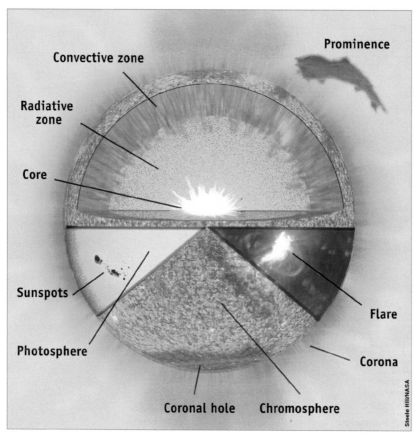

Plate 1. A cutaway showing the interior, and also features at the surface and in the atmosphere above. Image courtesy of Steele Hill, SOHO, ESA/NASA.

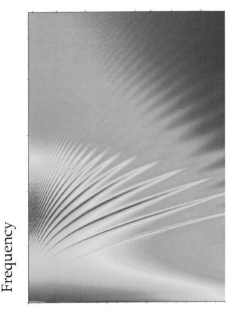

Frequency

Horizontal wavenumber /
angular degree

Plate 2. An $l\,\nu$ (or a $k\,\omega$) diagram made from data collected by the Michelson Doppler Imager (MDI) instrument on board the ESA/NASA SOHO spacecraft. The quality of this should be contrasted with the original diagrams shown in the final part of Chapter 4. Image courtesy of the MDI team, ESA/NASA.

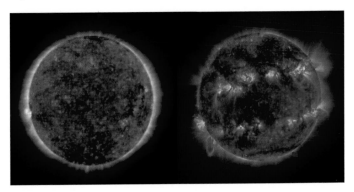

Plate 3. The Sun as observed at ultra-violet wavelengths, by the Extreme Ultraviolet Imaging Telescope (EIT) on board the SOHO spacecraft. The images show emission by iron atoms in the solar corona at temperatures of over one million degrees centigrade. The left-hand image was taken during a period of low solar activity, the right-hand image when the activity had begun its rise towards the peak level of the cycle. In the right-hand image, bands of strong magnetic field are wrapped in an east–west, toroidal manner around the Sun, at preferred bands of latitude (see text). These bands are where sunspots are found. Images courtesy of the ETT Consortium, ESA/NASA.

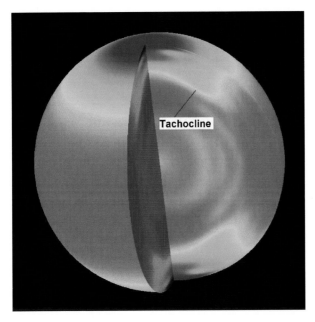

Plate 4. Three-dimensional cut-away showing the rotation rate in the solar interior, as derived from Michelson Doppler Imager (MDI) observations. Regions in red rotate more rapidly than those in blue. The tachocline mediates the transition from differential rotation in the outer layers to near solid-body rotation below. Rotation image courtesy of P.H. Scherrer and the MDI team, and ESA/NASA.

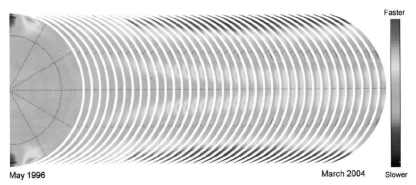

May 1996   March 2004   Slower

Plate 5. The 'torsional' oscillations in the outer layers of the solar interior, as revealed from the analysis of 72-day chunks of data collected by the Michelson Doppler Imager (MDI). The large-scale pattern of differential rotation seen at low activity has been removed from each set, leaving the much weaker torsional signal. Faster-than-average rotating regions are rendered in red, slower in blue (see coloured key). Over time, bands at moderate latitudes migrate towards the equator (regions in yellow, close to the equator), and a strong poleward-moving flow is present at high latitudes. Image courtesy of S. Vorontsov and collaborators.

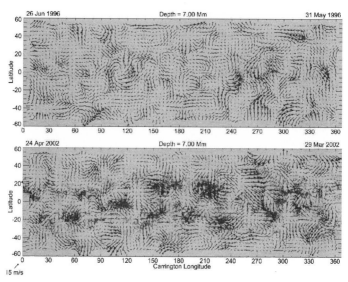

Plate 6. Plots of the solar sub-surface weather inferred from local helioseismic techniques (blue arrows). These are shown for a depth 7000 kilometres below the solar surface. Underlaid regions, in red and green, show locations of active regions of strong magnetic field. As the Sun rotates, different parts of the surface become accessible for observation. After about one month most of the near-surface material will have made a full rotation; analysis of these data allows a map to be made of the flows found over the surface during this time (as shown above, with regions towards the right from observations made at later times, when the rotation swept them into view). Image courtesy of D.A. Haber, B.W. Hindman, R.S. Bogart, M.J. Thompson and J. Toomre, from analysis of MDI data.

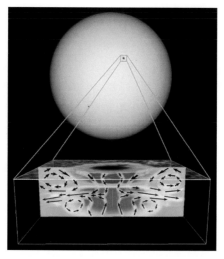

Plate 7. Variation of wave speed (high in red, low in blue) and flow of material (arrows) beneath a sunspot. Courtesy A.G. Kosovichev, the MDI team, ESA/NASA.

spectra do not discriminate between this type of useful periodicity, in the observed data itself, and that which arises from the absence of data, like the gaps each day from when the Sun sets. Gaps in single-site data therefore put structure into the frequency spectrum. New features, called sidebands, are introduced around the real mode peaks at multiples in frequency of one over 24 hours. This spacing corresponds to just over 11 micro-Hertz. You may recall, from Chapter 5, that 11 micro-Hertz is similar to the spacing between some of the low-degree mode peaks. Spectra are therefore likely to get even busier in appearance and harder to analyse properly. Aside from this periodic sideband component, the gap structure will also have a more random element. This puts extra noise into the background in the vicinity of the modes.

Both the sidebands and broadband noise are reduced in size when gaps in the data are minimised. The ideal is therefore complete, hundred-per-cent coverage – and no sidebands and extra noise.

So, to the wish list of lengthy, year-round observations had to be added the requirement the data be as near continuous as possible. Next, there was the impetus provided by the low-frequency modes. The early helioseismic spectra showed the observed power in the modes was strongly frequency dependent. With the accumulation of longer, higher-quality stretches of data more and more modes began to pop out of the background at low frequencies, modes that were successively lower overtones of their respective fundamentals.

The uncovered peaks carried with them the attractive property that they got narrower and narrower in frequency the further down the spectrum one went. The fact the peaks got narrower told the scientists the mode lifetimes were getting longer. This was a good thing, for it meant the frequencies could be determined to very high levels of precision – and also superior levels of accuracy, thanks to the peaks not overlapping as much.

Extended observational coverage would offer major improvements in the study of low-frequency modes. With more data it would be possible to beat down the background noise, allowing the narrow, low-frequency peaks to emerge.

Then there was the issue of solar variability. Even before evidence began to emerge that the frequencies of modes underwent systematic and measurable changes, correlated strongly with shifting levels of activity on the Sun, it was realised the modes might give invaluable insight into the origins of the solar activity cycle. This provided irrefutable scientific demand for long-term, year-on-year monitoring. Understanding the processes that underlie solar variability is important not only for solar and stellar theory but also for getting a better appreciation of the Sun–Earth connection, the most headline-grabbing aspect of the latter being the relationship of solar input to terrestrial climate change.

How could a year-on-year monitoring capability be established? A South Pole station would not permit year-round observations. In fact, the need for safe transportation to the site used by Jack Harvey, Tom Duvall and colleagues had limited its observing season to a 3½-month period from late October to early February each year.

There were two obvious approaches to the problem: either deploy a network of helioseismic observatories around the Earth (the cheaper option); or place a dedicated instrument on board a spacecraft with an unobstructed view of the Sun (the more expensive option). Although not originally intended for the purpose, the Active Cavity Radiometer Irradiance Monitor (ACRIM) on board NASA's Solar Maximum Mission satellite demonstrated in the early 1980s that the modes could be observed comfortably from space. The instrument was designed to monitor the radiation output of the Sun, but its observations were also of sufficient precision to allow the small fluctuations in intensity associated with the oscillations to be uncovered. It would be from analyses of these data that the first reports of activity-related changes to the modes were made. The need for a bespoke helioseismic, space-borne instrument – in particular one with the capability to resolve the Sun into many pixels, a capability ACRIM did not possess – was clearly indicated.

The timeline of development would, inevitably, see the ground-based networks established first. Anyone familiar with the tremendous effort needed to plan, build, test and launch a satellite will know the many years such an undertaking consumes. I shall begin with an overview of how ground-based network operations developed.

The Birmingham group was the first to establish a dedicated global network. The decision to place a good deal of its funding (and reputation) on the line by building and demonstrating the reliable operation of an automated station can be seen today as an important moment for helioseismology. This was one of the first automated astronomical telescopes anywhere, in any field.

Before 1979 and the birth of global helioseismology, George Isaak and his colleagues already had in mind the deployment of a permanent network. A good deal of their effort was devoted to trying to understand the 160-minute signal. Since the strength of the recorded signal was not greatly above the background noise, the Birmingham team began to seek funding to expand operations beyond its Tenerife-based instrument (christened the 'Mark I') in order to improve their observations. As early as 1976, George travelled around the United States looking for prospective sites and welcoming institutions. Patience was needed. In spite of its success, the Birmingham team had enough of a problem securing funding to continue operations at one site, let alone two or more. But by (in George's words) 'scrounging, scrimping and saving' the group was able to support additional observations in Pic-du-Midi (in 1978 and 1979) and Calar Alto in Spain (in 1980). The fact the team could show that a common pattern of 5-minute oscillation was present in data from more than one site was an important part of its 1979 global-mode discovery paper.

In 1981, things began to turn around. The group secured funding and an agreement to base an instrument at the Mees Observatory, 3000 metres above sea level on the Hawaiian island of Haleakala. In its first incarnation the instrument required daily attention, so operations were limited to those times of year when members of the Birmingham team could be on site. The small spectrometer received its sunlight via two mirrors. The kennel-like housing had to be rolled back each day to expose the mirrors, and equipment turned on and aligned.

Figure 7.2. At the Observatorio del Teide on Tenerife – the Birmingham Mark I instrument is housed in the pyramid building shown on the right, and fed sunlight by a system of mirrors (a cœlostat). Pico de Teide looms large in the background. Picture courtesy of S.J. Hale.

Two sites were now up and running – one on Haleakala, the other on Tenerife. On completion of his PhD thesis with the Birmingham group, Teo Roca Cortés had taken up a position at the Tenerife-based Instituto de Astrofísica de Canarias. There, he began to build up his own helio-seismology group. Teo and his colleagues continued to work with the Birmingham team and ran the main Birmingham instrument (housed permanently at Observatorio del Teide, Figure 7.2) and several others in collaboration.

Even with two sites operations were still a long way from anything resembling year-round, near-complete coverage. At this point the Birmingham group made an important policy decision that was to determine its course over the next two decades. Automation would be the key. It was felt that by developing automated systems and procedures it would be possible to deploy a dedicated global network – essentially copies of the demonstrator station – at a reasonable cost. The decision was quickly made that this new breed of instrument would get its sunlight in a different way. Whereas mirrors had fed the previous design, the new spectrometers would point directly at the Sun. To track the Sun, the instruments would be supported astronomical-style on an equatorial mount.

The team worked hard to get an instrument and mount ready for testing, on Haleakala, in the summer of 1983. The initial tests showed the spectrometer could track the Sun satisfactorily. With these promising results in hand, members of the team then turned their attention to the design of a new, modular plug-in instrument and the selection of a location for the first permanent new-breed site.

Four sites were considered, all in Western Australia. The first was the Woomera rocket range. It was from here that the UK made its one and only successful satellite launch in 1971. This sent the Prospero satellite into orbit on a Black Arrow launch vehicle. George concluded that Woomera was far too dusty. The other sites – at Learmonth, nearby Exmouth and Carnarvon – were all more than suitable. In the end, George plumped for Carnarvon.

The choice of Carnarvon was predicated largely on the lower temperatures found there. The team felt the more temperate climate offered better prospects for long-term operations. Less strain would be placed on the cooling units needed for the equipment.

NASA had used Carnarvon during its manned space programme to host a remote-tracking station. The big tracking dish was used to control missions passing overhead during the Mercury and Gemini programmes and as a communications relay during the Apollo programme. One of the astronauts assigned there in 1965, as a back-up capsule communicator for the Gemini 4 mission – on which Ed White made the first space walk by a US astronaut – described Carnarvon as a sleepy fishing village with just one hotel. When the Birmingham team arrived in 1984, the town still felt very isolated.

For this first visit a shakedown of the instrumentation was conducted, with the mount located on the flattest part of desert the team could find. At night the set-up was covered with a tent. Group member Roger New recalls vividly how, as is often the case with observational and experimental science, many things were learned the hard way. Only by running the actual set-up for several weeks did the team discover that certain types of cable connector did not take kindly to being swung around on the back of an instrument. Luck (or rather a lack of it) dictated the team had lots of the types of connectors that soon disintegrated, but few of those that held firm. Even though it was the Australian winter, high temperatures caused parts of the equipment to keep overheating. Mobile fridges (of the type usually intended to house a certain refreshing type of drink) were used to try to alleviate the problem. However, the difficulties that caused the most mirth (and embarrassment) concerned the mount itself.

An equatorial mount works by having its principal axis of rotation aligned parallel to the Earth's rotation axis. Since Carnarvon is at a latitude just under 25 degrees south of the equator, the principle axis of its mount needs to be tipped at the same angle above level ground when on site. Over the course of a day, the spectrometer tracks the Sun by being rotated around this axis at a rate that matches the Sun's motion across the sky. Servomechanisms are then used to fine-tune the guiding so that the instrument collects light from the whole of the visible disc as it tracks. The direction in which the mount needs to turn depends on whether it lies in the northern or southern hemisphere.

When it was first turned on, the Carnarvon servomechanism worked straight away – the system tracked the Sun. However, suspicions that something might be wrong began to grow. A constant, loud whirring indicated the servomechanism motor was working much harder than expected. It soon transpired the equatorial drive was turning in the wrong direction. The servos were therefore desperately attempting to counteract the nuisance of a drive that thought it was in the northern, not the southern, hemisphere (something the team had forgotten to swap over).

By the end of the summer the team members had ironed out many small problems and added significantly to their understanding of how a spectrometer–mount combination should be run. The results had been so encouraging that the decision was made to forget the original interim goal of semi-automation – they would go for full automation right away.

Before the practical details of this could be mapped out, a decision had to be made on how to permanently house the spectrometer and mount (volume equivalent to that of a small car) and the associated control computer and electronics. Funding problems were also looming.

Through his powers of persuasion, George managed to convince his then Head of Department in Birmingham, Professor Derek Colley, to provide a one-off, up-front payment of £10,000. This was regarded as an investment, much-needed funds to help George and his colleagues persuade the then Science Research Council that automation was feasible and money ought to be forthcoming to support a network. George agreed he would repay the university money on a timescale of five years. The network proved to be such a success that the debt was written off when the network was rolled out. Although £10,000 hardly strikes one as a large amount of money now, then it provided a much-needed boost. It allowed the group to push ahead with its plans for Carnarvon.

The initial pitches for housing the mount and instrument were: a 'dog-shed-like' structure, which would have cost about £200; and a 'supported roof on a rail track', which had an estimated cost of perhaps £2000. In economic terms, the dog shed looked the attractive option. However, the relative merits of the two were thrown up in the air when Clive McLeod made the observation that 'astronomers use domes'! Other members of the team recall this obvious solution had not dawned on them until then. A dome was probably the most expensive option, but this option would allow the team to use a well-tested and reliable commercial product.

With the decision to use a dome made, procedures, hardware and software needed for automation were developed. At the heart of the system was a Hewlett Packard desktop computer. This ran a control program to calculate the co-ordinates of the Sun at the local site. Output from this could have been used to control directly the mount servomechanism. However, the team was worried the computer would not be able to interface quickly enough to perform this task effectively. Instead, the computer would control a servo that externally matched the Sun's position to the position the computer was commanding.

A separate, and less satisfactory, system was tried as a means of controlling the position of the dome's open aperture. In essence, the dome was slaved to the mount. Detectors in each corner of the aperture measured the signal from an infrared emitter at the centre of the mount. The idea was that the dome would move to keep the signals balanced and the aperture aligned in the right place, centred on and facing the mount and spectrometer. The aperture system needed to be tested in Birmingham – but the dome had already been shipped directly to the site in Carnarvon. The task of getting the system running had fallen to Clive McLeod. He remembers running around with Roger New on the roof of the Physics Department, holding a large frame with sensors at the corners which was designed to mimic the aperture of the dome. When the system was installed in Carnarvon it worked fairly well. However, the team was never entirely happy with it – for one thing the Sun is a conspicuous infrared source and gave rise to confusing signals. Eventually, the team opted for the better approach of fixing home-made encoders to the commercial dome. These allowed the position of the dome to be accurately recorded and relayed back to the computer.

Local contractors put down a concrete base to support the dome in Carnarvon during the middle of 1985. A central plinth was also added on which the mount and instrument would stand. Members of the Birmingham team then travelled to the site to assemble the commercially purchased dome over the base and plinth (Figure 7.3).

Marine plywood and corrugated iron formed the lower, cylindrical structure. This served as a ground-floor area to house the computer and electronics. Flooring was then added on top of the plinth and lower structure. This upper level housed the instrument. Finally the dome was added on top.

In all, it took a few weeks to put everything together, during which time the scientists benefited greatly from the help of a local rigger called Basil Adams. Basil put up scaffolding around the growing dome to help

(a)

(c)

(b)

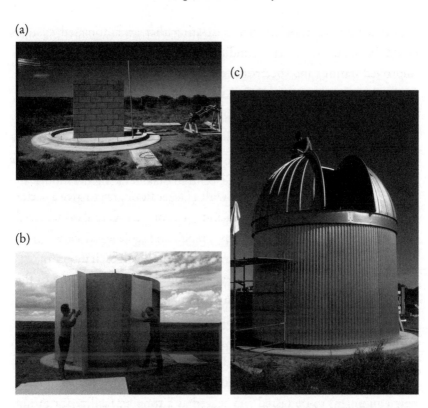

Figure 7.3 (a–c). Carnarvon – the Australian outback, 1985, setting up the first automated helioseismic observatory. (a) The central plinth, and the instrument on its equatorial mount. (b) Neil Pearson and Roger New construct the dome from a 'kit'. (c) Almost complete! Picture courtesy of C.P. McLeod.

with the construction. This was cannibalised from the support structure of one of the nearby defunct antenna dishes, a task that demanded some care lest the big dish collapse in a heap. Basil nicknamed Clive McLeod, who was on the trip, 'Lightning': when he hammered in nails Clive would never strike the same place twice.

Over the coming months the system repaid the group members for all their toil with high-quality data and a pleasing record of reliability. The goal had been achieved of demonstrating that an automated system could be made to work. Funding to deploy a global network of improved stations and spectrometers was soon secured.

At the same time the existing site on Haleakala was modified so it could run in an automated manner. The same instrument is still collecting data almost two decades later.

The principal changes made to later automated stations largely concerned ease of access. The support buildings would instead be permanent brick-built constructions with a larger floor area to give a better working environment. The mount and spectrometer were also increased in size. This made it easier to remove, repair and replace individual parts and allowed more than one instrument to be hung on each mount.

The first new-type observatory was built on the roof of one of the Physics Department buildings in Birmingham, in 1988. Because the Poynting Building is officially listed, formal planning permission was required.

The Birmingham dome served as a demonstrator of the improved system, copies of which were rolled out to three new sites. It also allowed new equipment to be tested and served as a training facility for group personnel.

A systematic pattern of deployment saw stations established in: the Karoo desert in South Africa, in 1990 (at the South African Astronomical Observatory); the Andes mountains in Chile, in 1991 (at the Las Campanas Observatory); and Narrabri, in New South Wales, in 1992 (at the Australia National Telescope Facility).

The year 1992 marked a final change to the network. The group moved its Hawaiian instrument (they could not afford the high site fees on Haleakala) to the Mount Wilson Observatory, where it has since been housed in Ed Rhodes's 60-foot solar tower (and looked after by Ed and his personnel). The network thereby reached its current six-station configuration. Over a decade on, the network – now called the Birmingham Solar Oscillations Network (BiSON) – continues to collect high-quality data, achieving annual coverage in the low to mid eighty per cent range.

The BiSON arose out of the pioneering work of George Isaak and his colleagues. Installation d'un Réseau International de Sismologie Solaire (IRIS) was the natural follow-on to the work of Eric Fossat and Gérard Grec. Proposals for a network were made in 1983 (with Bernard Gelly) and funding received for a multi-station programme the following year.

In planning the network, Eric decided to adopt a different working philosophy from BiSON's. He feared automation would prove too expensive to develop and maintain, given the limited level of government support available to him. This meant local help would be needed to run the instruments at each of his chosen sites. Although the sites were selected to give as even a distribution in longitude as possible, Eric also had to find places where the locals were not merely willing but positively enthusiastic about contributing to the project. Aside from the political dimension, folk had to be found with enough motivation to run the instruments day after day, no matter what the time of year.

A robust network prototype instrument – fed sunlight by a two-mirror system – was run successfully at the La Silla Observatory in Chile from the latter half of 1986 to the middle of 1987. The roll-out of stations began the same year. Because of funding problems the team was forced to operate on a shoestring budget. The network continued its operations until 2000, which gave a neat ending to the project – available data stretched back over a full 11-year cycle of activity. The IRIS dataset stands as another valuable resource for the helioseismology community at large, and a careful re-processing of the data has recently been performed.

Eric singles out two of the lesser-known sites as having been particularly successful – those at Kumbel in Uzbekistan and l'Oukaimeden in Morocco. Their success was very much down to the enthusiasm and dedication of the local staff. Without a project like IRIS, the locals might never have had the opportunity to participate in cutting-edge astronomical research. Local students are still running the Moroccan instrument some four years after IRIS officially came to an end.

Uzbekistan was never the easiest place to visit. While it was part of the old Soviet Union extensive paperwork was required to travel there. Shipping and installation of equipment were particularly difficult. After the fall of the Soviet system, political unrest made it an uncertain place to work. However, the expertise and dedication of the local staff made it a key contributor to the science that arose out of IRIS.

Both BiSON and IRIS were established to observe the core-penetrating, low-angular-degree modes. But the first major ground-based network with the capability to detect modes at higher degrees was the Global Oscillations Network Group (GONG). This network is run from Tucson, in Arizona, by a group of scientists led by John Leibacher who are based at the National Solar Observatory. Each of the GONG instruments observes a very rich spectrum of modes, from the lowest angular degrees up to $l \approx 1000$, by making resolved observations of the visible solar disc.

GONG can trace its heritage, in part, to discussions in the United States in the early 1980s. At the time, proposals for a joint European Space Agency (ESA) and NASA mission to put a satellite in a polar orbit around the Sun were being considered. This orbit made the International Solar Polar Mission (ISPM) novel. By passing well out of the ecliptic plane the spacecraft would be able to study in detail the high-latitude regions of the Sun. In 1981 the Reagan administration pulled funding, forcing NASA to cancel its contribution to the mission. The payload of ISPM was first shifted to the NASA Solar Interplanetary Satellite. The definition study proposed placing this satellite in a drifting orbit behind the Sun. This offered the possibility of a stereo view by combining observations with those of a ground-based network. At this point there was some discussion about the Solar Interplanetary Satellite and the network being given a helioseismic capability. However, the satellite was never approved. The original aims of ISPM were later fulfilled by the highly successful Ulysses mission, but with no helioseismology instruments on board. As we shall see later, a helioseismic package would fly on the Solar and Heliospheric Observatory (SOHO) spacecraft in late 1995.

The concept of a US-run network was in tune with part of the National Solar Observatory's mission – then only recently formed from the merger of the Kitt Peak and Sacramento Peak observatories – this being to provide state-of-the-art observing facilities for the solar physics community. Jack Zirker, the National Solar Observatory's director, knew he had several eminent helioseismologists on his staff. At a meeting he called at the McMath–Pierce Telescope it was decided to put together a formal bid for a global network of helioseismic observatories.

John Leibacher would head GONG's team as the Project Scientist. Jack Harvey was the obvious choice to head the instrumentation effort and he began to look at the various options. It was decided the instruments should initially be made capable of observing modes up to angular degree 150. This meant the Sun had to be resolved into patches each with a length on the side about one-hundredth the size of the radius of the Sun. A detector split into 250 by 250 pixels could do the job. A technique then had to be selected to measure the Doppler shift of each patch. Four were considered: the Fourier Tachometer approach, used by Tim Brown; the magneto-optical resonance filter developed by Alessandro Cacciani (then also being developed for use in Ed Rhodes's Mount Wilson programme); a Fabry-Perot interferometer, advocated by Dave Rust; and a filter similar to the one Ken Libbrecht would go on to use so successfully at Big Bear.

Frank Hill looked after data collection and analysis. His first responsibility was to organise the collection of survey data from several candidate sites around the globe. The analysis of these data would be used in selecting the final sites. Simulations performed by Frank, and Gordon Newkirk, suggested at least six sites were needed to guarantee high-quality, year-round coverage. In the event, six stations were deployed.

Throughout the initial study phase the GONG team also had valuable input from a scientific advisory committee comprising Juri Toomre, Peter Gilman, Bob Noyes, Alan Title and Roger Ulrich. The team was encouraged by the pioneering work of the Birmingham group in demonstrating that remote automation was feasible.

The result of the study phase was a detailed proposal to the National Science Foundation in 1985 for funding to deploy a six-station network. This was considered to be in competition with proposals for an infrared telescope. Jack was serving on an advisory committee at which the GONG application was reviewed. Although he was not allowed to pass comment on the proposal, Jack was not asked to leave the room during the discussions. He remembers sitting quietly in the corner, offering a furrowed brow to anyone who said something he did not agree with.

To the delight of the National Solar Observatory and the helio-seismology community, GONG was successful. The proposal sought funding to run initially for three years, but noted, 'the network could be continued for the full 11-year solar cycle after appropriate review'.[1] Even though the bid received about half the funds requested, this proved sufficient to roll out a full network.

One of the biggest worries of the team had been the huge chunk of the budget that had to be set aside to save the data. The main storage medium was then nine-track tape. Had these tapes turned out to be the only means of storage available, Frank Hill estimates the costs just to hold the data would have soared to tens of millions of dollars. Frank says everyone crossed their fingers in the hope that advances in computer technology would save their bacon (which they did).

On the instrument side, after a careful comparison of the four options, Jack followed the advice of a committee of outside experts and opted for the Fourier Tachometer approach.[2]

The late 1980s were to prove a busy time for Jack. He and Tom Duvall went back to the South Pole to make a new series of observations, with Martin Pomerantz and Stuart Jefferies. The development of the GONG prototype instrument was also coming along. There were lots of

---

[1] 'The Global Oscillation Network Group Project: A Proposal to Study the Solar Interior by Measuring Global Oscillations with a World-Wide Network of Instruments', Global Oscillations Network Group, National Solar Observatory, National Optical Astronomy Observatories, 1985, p. 10.

[2] At an important meeting in May 1985 three outside experts advised the team to go with the Fourier Tachometer approach but also warned it would be very difficult.

Figure 7.4. Lined up and ready to go – the six GONG observing facilities have one final get-together before heading off to their new homes around the globe. Picture courtesy of GONG/NSO/AURA/NSF.

problems to be worked out. Even though funding was sluggish the slow release of money proved to be a blessing in disguise in that it gave the team more time to get the prototype working properly.

In early 1991 the final selection of six sites from fifteen was made (Figure 7.4). Those chosen were: Big Bear Solar Observatory, in California; Learmonth Solar Observatory, in Western Australia; Udaipur Solar Observatory, in India; Observatorio del Teide, in Tenerife (hosted by Teo, Pere Pallé and colleagues); Cerro Tololo Interamerican Observatory, in Chile; and Mauna Loa Observatory, in Hawaii. All of the sites would be partners; five were already thriving observatories with active programmes.

Each instrument would be fed sunlight by mirrors. The instrument, electronics, computers – in fact each whole station – was housed inside a commercial shipping crate to be safely shipped to its destination. As is the nature of instrumentation development, what was implemented bore little relation, in some respects, to what had originally been proposed. However, after several years of careful research and testing, the deployment of the network began in 1994; the various sites saw first light between February and September 1995.

GONG has proven to be a wonderful success. The high-quality data and excellent science results meant the decision to extend operations beyond the original 3-year lifetime was an easy one to make. In its original configuration – GONG 'Classic' – the full, potential capability of the instrumentation was not exploited because of limitations imposed by 1980s TV technology. This is no longer the case. The instruments were all updated, with higher-resolution camera detectors, not long after the turn of the millennium; 'GONG++' can now resolve even finer scales on the solar surface than was possible with GONG Classic. GONG is part of the permanent programme of the NSO and has a bright future ahead of it.

The mid 1990s produced an explosion in the number of research papers and scientists working in helioseismology. It is no accident this coincided with the increase in the amount of data available. On the ground-based side, GONG had just come on line and was beginning to produce scientific results covering a vast range of topics. BiSON had put a few years of full, six-station coverage under its belt, and this was beginning to reap benefit, particularly for detection and study of the low-frequency modes. Meanwhile IRIS was entering its most productive data phase.

In 1994, Steve Tomczyk's LOWL instrument began making observations in Hawaii.[3] LOWL provided high-quality data on modes with angular degrees from zero all the way up to about one hundred. Steve had cut his teeth working with Alessandro Cacciani to develop an instrument for Ed Rhodes's programme at Mount Wilson (Figure 7.5). The Mount Wilson system was designed to make observations of high-degree modes; it was fed sunlight by the 60-foot tower at the observatory. The system began

---

[3] This became a two-station network – Experiment for Co-ordinated Helioseismic Observations (ECHO) – at the end of 1999 when a new instrument was added at the Observatorio del Teide. The instrument was named LOWL because it was optimised to measure the deeply penetrating LOW-degree (L) modes.

Figure 7.5. Ed Rhodes (seated at front) with staff from the 60-foot and 150-foot towers at the Mount Wilson observatory. From left to right: Pam Gilman, Johan Boyden, Natasha Johnson, Jim Frazier, Ed, Maynard Clark, Kirk Palmer and Steve Padilla. Picture courtesy of E. J. Rhodes, Jr.

regular operations in 1987. In 1996, a similar set-up, overseen by Leonid Didkovsky, was installed at the Crimean Astrophysical Observatory, thereby establishing the two-site High-Degree Helioseismology Network (HiDHN). That same year Dean-Yi Chou's Taiwan Oscillation Network (TON) added a fourth site. TON was also designed to study modes with very high angular degrees; after trial observations in 1992 its first instrument proper was deployed to the Observatorio del Teide in 1993.[4] Subsequently, instruments were added in Beijing, Big Bear in California and Tashkent in Uzbekistan.

However, the big observational landmark of 1996 came with the start of operations of the ESA–NASA spacecraft SOHO, which carried four helioseismology instruments into space and put them at a vantage point that offered a continuous, unobstructed view of the Sun.

[4] Teo Roca Cortés, Pere Pallé and their colleagues have hosted instruments belonging to BiSON, IRIS, TON, GONG, ECHO and the Luminosity Oscillations Imager (LOI) prototype.

SOHO can trace its heritage back to proposals made in the 1970s. The spacecraft ended up essentially with two packages of instruments – a helioseismology package, and a suite dedicated to observing the chromosphere, corona and solar wind. The forerunner of the coronal package came first – a proposal for an instrument called the Grazing Incidence Solar Telescope, which was intended to fly as part of a Spacelab payload on the Space Shuttle in the 1980s. Although this instrument eventually fell foul of the political fallout from the cancellation of the ISPM, its scientific objectives were largely carried over to SOHO.

The first predecessor of the latter's helioseismology package was the wonderfully named Dual Spectral Irradiance and Solar Constant Orbiter (DISCO). When originally proposed by Roger Bonnet and collaborators, in July 1980, the instrument did not have a helioseismic capability. It was at about this time that the implications of the global-mode observations of the Birmingham and Nice groups were being taken on board by the community. The decision was quickly made to add this capability to the proposal, and the late Philipe Delache began pushing the revised mission. This was now to include a Sun-as-a-star Doppler instrument, a miniaturised version of that used at the South Pole by Eric Fossat and his colleagues. The technical specification of the original instrument package was also modified to allow DISCO to detect oscillations in intensity as well. Finally, Bonnet and Delache decided the instrument should be located at the 'L1' Lagrangian point between the Sun and the Earth. This is a neutral point at which the forces of gravitational attraction of the two bodies balance to give the centrifugal force needed for an object placed there to orbit the Sun with the same period as the Earth.[5] It is located about 1½ million kilometres from the Earth.

---

[5] The L1 location is unstable in the Sun–Earth direction. SOHO orbits L1 every six months in a plane perpendicular to the Sun–Earth line. This 'halo' orbit carries communications benefits. When a spacecraft is aligned precisely on the Sun–Earth line, the Sun can cause interference with radio communications, since it is a prodigious emitter at radio frequencies.

All three principal elements of DISCO were preserved on SOHO when it was finally launched in the mid 1990s. First, the South-Pole-inspired instrument had evolved into the Global Oscillations at Low Frequency (GOLF) spectrometer. Second, the package designed to measure the oscillations in intensity flew as part of the Variability of Solar Iradiance and Gravity Oscillations (VIRGO) package. And, third, SOHO was launched to L1. However, before this, DISCO had first to run its course.

In July 1982 a call for mission proposals was released by the ESA Director of Science to coincide with a new planning cycle for prospective satellite missions. Proposals would of course be forthcoming from the solar physics community. However, the scientists were aware of a potential problem – that the deadline would fall before the solar community knew whether DISCO, which was already under consideration, had been selected for future development. The feeling was that pitches for helioseismology instruments, made as part of the new call, might be viewed as being in competition with DISCO rather than complementary. If the two had been regarded as mutually exclusive, any new proposal might have had the potential to kill DISCO in its tracks. Despite efforts by several leading figures, the deadline remained fixed.

Discussions in the community highlighted the need for a mission that would combine the best elements of DISCO and the Grazing Incidence Solar Telescope. A new satellite was envisaged which would be launched into a low Earth orbit. During these early stages a careful balancing act had to be struck. There was an obvious need to look at a potential helioseismic package for the new mission. However, this could not be trumpeted too loudly lest it compromise DISCO's chances. This is why most of the initial work concentrated on the requirements of the instruments designed to study the solar corona. (Also, many of the solar scientists who were first involved had their main interests in atmospheric and coronal physics.) The mission was soon dubbed the 'Solar High-Resolution Observatory', or 'SOHO' for short.

The DISCO proposal survived until early 1983, when it lost out to the Infrared Solar Observatory. However, the ongoing SOHO study allowed for an immediate resurrection of DISCO's main helioseismic goals. A helioseismology package was thus formally added to the SOHO payload and the meaning of the SOHO acronym was changed to the 'Solar and Heliospheric Observatory'. The orbit was also altered to take it to L1. This would give the uninterrupted coverage desired by the helioseismology community.

Throughout this period similar studies of the desirability of a long-term, space-borne mission had been ongoing in the US. Formal activity dated back to 1978, when NASA set up a Space Working Group to look at the options to follow up the ACRIM instrument on the Solar Maximum Mission satellite.

In the US, proposals for large science projects benefit greatly from exposure in National Academy of Sciences (National Research Council) reports. This body is regularly asked by government funding agencies to commission reports to oversee certain fields and help guide the use of public money. In the early 1980s helioseismology received the blessing of several such reports. One of these, produced by a Working Group on Solar Physics (led by Art Walker), endorsed not only a ground-based option (which led to GONG), but also a space mission. Jack Harvey and Roger Ulrich were both members of this group.

NASA soon commissioned a second Space Working Group to consider in detail the observational requirements for helioseismology. This group, chaired latterly by Bob Noyes and co-chaired by Ed Rhodes, reported its findings in 1984. Its membership read like a who's who of helioseismology and included twenty-five names from the US and Europe.

The report made two important recommendations. First, with regard to a possible space mission, it encouraged NASA to develop an instrument capable of making high-resolution observations of the solar oscillations by the Doppler method. It recommended this be flown to the L1 point. Second, the group stressed the importance of having complementary observations available from ground-based observatories or networks of stations.

The Committee (of the Space Science Board) on Solar and Space Physics also issued a number of reports during this period. Further endorsements of helioseismology were included in the likes of 'A Strategy for the Explorer Program for Solar and Space Physics' (1984) and 'An Implementation Plan for Priorities in Solar-System Space Physics' (1985).

In 1983, NASA, ESA and the Japanese Institute for Space and Astronautical Science (ISAS) came together to rationalise their planned efforts in the solar and terrestrial physics area. The outcome of extensive discussion was the blueprint for an International Solar–Terrestrial Physics (ISTP) programme of satellite missions. A Joint Planning Group was set up to oversee the development of the ISTP. Independently of this, ESA had entered negotiations with NASA to make SOHO a joint mission.

A stimulus for the ESA–NASA discussions had been the realisation that the combined cost of SOHO and another planned mission would exceed the available ESA budget. NASA agreed to provide the launcher, operations centre and some smaller sub-systems for SOHO and to finance some of the science instruments.

With the SOHO project now up and running, a joint science advisory group was formed. The NASA contingent – which included Roger Ulrich and Ed Rhodes – pushed strongly for the capabilities of the helioseismic package to be extended beyond those of DISCO, which had been intended to make Sun-as-a-star observations only (albeit in both intensity and velocity). A high-resolution instrument was therefore added. This would fly on SOHO as the Michelson Doppler Imager (MDI). An imaging instrument was also added to the VIRGO package – this flew as the Luminosity Oscillations Imager (LOI).

After SOHO had been selected to fly, a joint ESA and NASA 'Announcement of Opportunity' was made to the community in March

Figure 7.6. The ESA/NASA SOHO spacecraft. Picture courtesy of ESA/NASA.

1987. This called for proposals to design and build the various instruments on the spacecraft (Figure 7.6).

Two competing proposals were made (from what had originally been one consortium) for the Sun-as-a-star Doppler instrument. The GOLF consortium, led initially by Luc Damé and later Alan Gabriel, were successful; they brought on board members from the team who lost out. GOLF was to be developed by a consortium comprising four institutes in France[6] and the Instituto de Astrofísica de Canarias, in Tenerife. It could draw on the considerable instrumental expertise and long-standing experience of the Tenerife and Nice groups.

Swiss scientist Claus Fröhlich had already led several teams that had designed and constructed space instruments for measuring the Sun's

[6] Institut d'Astrophysique Spatiale at Orsay, Paris; Service d'Astrophysique of the Commissariat à l'Energie Atomique at Saclay, Paris; Observatoire de l'Université Bordeaux 1; and Observatoire de la Côte d'Azur, in Nice. Roger Ulrich also served on the main Science Committee.

radiation output. His VIRGO consortium took on the task for the SOHO mission. The VIRGO package would contain instruments not only for monitoring the solar constant but also for detecting the oscillations in intensity. Two types of oscillation monitor were provided – in addition to the imaging capability given by the LOI, the VIRGO Sun photometers would do a Sun-as-a-star job.

The team responsible for the high-resolution Doppler instrument was initially not in the running, simply because it did not yet exist as a team. Phil Scherrer had been given the responsibility of chairing a sub-committee for one of the NASA working groups. This sub-committee was charged to look at the needs of the high-resolution instrument, for which two competing proposals were on the table. One was from NASA's Jet Propulsion Laboratory, which proposed using the magneto-optical filter technology employed at Mount Wilson. The other was led by Dave Rust, of the National Solar Observatory, and used an instrument called the Stable Solar Analyser (based on a widely used type of spectrometer called a Fabry–Perot filter). Phil's committee, comprising as it did many of the key scientists in the field, ended up making its own proposals for the instrument and was eventually selected to build it. The flight instrument for the Solar Oscillations Investigation programme flew as the MDI.

Phil notes that his proposal had an interesting advantage over the other two. It called for the use of an observational technique that had not been fully tested. The other proposals both had working prototypes. It might seem, therefore, that the latter were in a more advantageous position. However, working instruments have known niggles – those of the MDI had yet to see the light of day.

Although the instruments being developed by the various consortia had widely differing observational requirements, they had in common the basic, and very stringent, constraints imposed by any space mission.

First, there would be well-defined limitations on the mass and volume of the instruments. The size of a ground-based instrument is largely limited by practical engineering requirements and cost. For example, the later BiSON instruments had to be light enough to be supported on an equatorial mount. In turn, the mount had to be designed to fit inside a modestly sized dome. The larger and more complex the construction, the greater would be the cost. On the other hand an instrument and mount too compact would present difficulties for ease of access during testing and repair. Indeed, the BiSON team found that its original spectrometer–mount combination in Carnarvon was too compact to be fully serviceable. Later designs were therefore altered accordingly.

In the case of a space instrument mass and volume mean everything. The payload capacity and fairing[7] size of the launch vehicle fix an upper limit on the total mass and volume of the satellite. The potential mass of each instrument is in turn dictated by the number and nature of the other instruments on the spacecraft and the overall needs of the mission. The proposed length of the mission has an important impact – there must, for example, be enough fuel on board to keep the spacecraft in a stable position.

The satellite and its instruments all need power. On SOHO this was to be provided by two large arrays of solar cells. The size of these dictated the total power budget. This in turn dictated the budgets for each of the instruments.

Careful design and modelling of the thermal properties of the instruments were needed to ensure these instruments worked effectively in the harsh environment of space. Things are not quite as harsh as one might imagine, in that the spacecraft (on which the instruments are mounted) can control the thermal environment to some extent. On SOHO, the boundary conditions called for equipment to operate in an environment where the temperature would range anywhere from 0 to 40 °C.

On a spacecraft heat is exchanged by conduction and radiation – convection does not occur in vacuum. Radiation emissions can come from the equipment on the satellite or from external sources like the Sun. The resulting heating can, for example, affect the properties of lenses,

---

[7] The fairing is the aerodynamic sheath that covers the payload at the top of the launch vehicle.

Horizontal wavenumber/
angular degree

Plate 2. An *l–ν* (or a *k–ω*) diagram made from data collected by the Michelson Doppler Imager instrument (MDI) on board the ESA/NASA SOHO space-craft. The quality of this should be contrasted with the original diagrams shown in the final part of Chapter 4. See the plate section in the middle of this book for a full colour version. Image courtesy of the MDI team, ESA/NASA.

and cosmic rays can interfere with the operation of the instrument by triggering spurious events in the electronics. All of these factors must be taken into account.

The instruments then had to be robust enough to withstand the rigours of launch, during which they could be subjected to loads of up to twelve times the force of gravity. The quality required of the components, and the level of reliability, would also need to be higher than would be acceptable for a ground-based instrument. This is because the instruments would have to work without the safety net offered by regular maintenance.

Unsurprisingly, these various constraints carry with them a huge financial cost. Spacecraft do not come cheaply.

A carefully defined series of steps needs to be taken to get from a successful instrument proposal on paper to a real instrument ready to fly. The LOI instrument, part of the VIRGO package on SOHO, is a nice case in point. It was designed and built by scientists, engineers and technicians based at the European Space and Technology Centre (ESTEC) in The Netherlands, one of ESA's principal sites. The instrument was designed, by Bo Andersen, in the knowledge that it would have to weigh less than two-and-a-half bags of sugar and occupy a volume no larger than a small tool bag.

The instrument was designed to detect the oscillations by observation of tiny changes in the intensity of the sunlight. It would measure the variations found in a narrow wavelength band of visible light and was given a modest imaging capability to allow it to observe modes of angular degree up to $l = 6$. It remains the only dedicated helioseismic intensity instrument with an imaging capability. Because the intensity variations associated with the modes are so small – they are at the level of a few parts in a million – that unwanted fluctuations arising from the intervening terrestrial atmosphere seriously compromise the quality of ground-based data of this type. Getting above the atmosphere is therefore a must for serious intensity observations.

When Bo moved on from ESTEC in the late 1980s, the newly appointed Thierry Appourchaux took over responsibility for the instrument. After he had finalised the overall design and that of the individual components, construction of the various elements began. For Thierry and his colleagues this meant first having to look for suitable contractors to make some of the parts. It took about two years to turn design drawings into working kit that could be tested. Only after each part had demonstrated it would work to design standards and tolerances could the team turn to the detailed design, development and construction of the basic structure that would hold the various parts together. Careful planning was needed for how each part joined to the others mechanically, thermally and electrically.

Not everything went smoothly. For the first integration test of the optical set-up a structural model was used in place of the main mirror in the instrument box. This dummy mirror was built to have the same mass, size and shape as the actual mirror. It allowed the team to check that all parts of the optical system that guided sunlight through the instrument fitted snugly together – without the need to handle the real thing.

When the dummy was finally put in place everyone's heart sank at the sound of a sickening 'ping'. The dummy had snapped. Although anxiety was tempered somewhat by the fact the real mirror was safe, the box was clearly not big enough for the mirror. The team checked and re-checked the design. On paper the box was the right size, but it turned out it had not been built to the required dimensions.

The guiding system also had to be put through its paces. It had to keep an image of the Sun centred on the instrument's silicon photodiode detector (consisting of twelve unusually shaped pixels). Part of the system comprised special mounting mechanisms called gimbals. Gimbals are designed and built so that they fit together very precisely, with absolutely no margin for error. Even the tiniest amount of friction can make assembling them a real headache. The first LOI gimbal fitted together perfectly in the workshop – but the next (destined for the first integrated model) would not. After spending most of a weekend in a swimming pool mulling over the problem in his head (so his wife did not suspect he was working) Thierry managed to work out a reliable procedure for mounting and aligning the system.

With all of the components now tested, the next phase of development was to put the complete assembly through its paces – the so-called 'instrument-level' tests. In all, five instruments were built: structural and electrical models, then a Qualification Model, a Flight Spare and a Flight Model. The Qualification Model was the first of these to be put through the wringer.

It was shaken violently – on special shakers at the ESTEC Test Centre – to check the instrument would be up to handling the extreme conditions of launch. The initial test of the Qualification Model failed because screws on the guiding assembly had not been locked in place by glue. After the screws had been secured, the Qualification Model passed with flying colours.

The electrical and electronic equipment had to be thoroughly checked. Detailed measurements were made to ensure the systems of the LOI would not interfere with the other instruments, or the spacecraft itself. The measurements were conducted in the ESTEC Test Chamber – essentially a huge Faraday cage that isolates anything inside from electromagnetic emissions on the outside.

The instrument was then subjected to the conditions of space in a thermal vacuum chamber. The Large Space Simulator at ESTEC is over 2000 cubic metres in size. Stainless-steel shrouds in the chamber can be dropped to temperatures as low as –170 °C. And a special Sun Simulator can fire a 6-metre-diameter beam of light at the instruments to match the full intensity of the Sun.

After the Qualification Model passed all its tests, the Flight Spare and Model were allowed to take things a little easier in the acceptance-level tests (which they passed too). Ordinarily, the Qualification Model would now have done its job. However, the LOI team took advantage of the fact that the Qualification Model was a fully functioning instrument. It was shipped to Tenerife and installed at Observatorio del Teide in May 1994, where it remained, collecting data, until the middle of 2000. Thierry and his colleagues were aware that the intervening terrestrial atmosphere would compromise the performance of the Qualification Model, but running it gave them an ideal opportunity to test the operation of the instrument design before the launch of the Flight Model. The Qualification Model duly became the first instrument to detect, in intensity observations from the ground,[8] the effects of rotation on the low-degree modes. Not only did the Qualification Model survive the vibration test twice, but over the 6-year period it was based at Izaña the instrument did not have a single failure. That is what 'space quality' can do for you.

Thierry and his colleagues had one more problem to tackle prior to launch; one that had the potential to turn into a major crisis. In January 1995, with less than a year to go to launch, some of the components forming part of the guiding mechanism stopped working properly.

The components were incredibly fiddly devices called piezo-electric actuators. The sizes of these pieces could be finely controlled by the application of voltages of different sizes. It was in this manner that the actuators were used to control the tilt of a mirror resting on them in the instrument. During operation, up to 700 volts had to be applied to the actuators. For some, as yet unknown, reason the system on the Flight Model was now producing high-voltage arcs in the air in its immediate vicinity.

The actuators were made up of several hundred electrodes, which had to be soldered together. An investigation soon uncovered that problems with some of the soldering joints were giving rise to the arcing. This part of the system had to be completely re-designed and brand new parts were delivered that were free of the problem. The team had dodged one bullet. However, several months had passed since the problem had appeared and there was no longer enough time to replace the faulty actuators on the Flight Model and install, test and calibrate a new system.

---

[8] The first ground-based intensity observations of modes were made in 1984 by a multi-channel photometer built at ESTEC and operated at Observatorio del Teide by Antonio Jiménez, Pere Pallé, Teo Roca Cortés, Vicente Domingo and Sylvain Korzennik.

Step forward the Flight Spare. It could be re-jigged in time and would therefore now become the new Flight Model designate.

The Flight Model (née Spare) was integrated into VIRGO and launched from Cape Canaveral as part of the SOHO spacecraft by an Atlas-Centaur vehicle on 2 December 1995. The first major hurdle for the LOI came seventeen days later.

It is Christmas Eve 1995. Thierry has flu and a temperature of 40 °C. At home, he awaits a telephone call from the NASA mission headquarters. Vicente Domingo makes the call – Thierry discovers the main systems of the LOI are operating as expected and that, crucially, the protective cover at the front of the instrument has opened and closed on command. Since this prevents sunlight from entering the instrument when closed – the configuration in which it will remain during the coast to L1 – its successful operation means he goes to bed a happy man.

17 January 1996. Udo Telljohann, the principal electrical engineer, is taking care of the LOI on the day the cover is to be opened permanently so the instrument can begin its extended programme of observations. But when the command is sent the cover fails to open. Repeated attempts give the same result. Would over seven years of hard work end like this – a faulty cover, no sunlight and no data?

It goes without saying that Thierry and his colleagues were not going to lie down and accept failure without a fight. First, they had to find a likely explanation for the failure. After looking long and hard at the physical properties of the cover, they found that under certain circumstances an unexpected mode of failure could result. This arose when the voltage pulse sent to the motor to open the cover persisted in time for longer than expected. The cover was then opened with too much force and ended up bouncing back into the closed position – like pushing a swing door open and having it come right back in your face.

This scenario seemed plausible, but finding a solution to it was less than simple. If an ill-timed pulse length were indeed the culprit, the obvious solution would be to re-set the length of the pulse to something shorter and try again. However, VIRGO had no on-board computer. This meant the various commands being executed by the electronics and mechanical elements could not be re-programmed.

It was during a coffee break that one of Thierry's scientific colleagues in the then Solar System Division at ESTEC hit upon a solution. Trevor Sanderson noted that, although VIRGO was not programmable, the SOHO platform was. VIRGO received its power from SOHO. What if the spacecraft were to remove power partway through a pulse? This might provide a way to shorten the pulse and prevent the cover from bouncing shut.

Team members reconvened at Davos, in Switzerland, where the VIRGO Flight Spare was located. They performed a series of tests on the cover, which showed it was behaving in many respects as Thierry had predicted. Thierry's video camera provided valuable data on the cover's behaviour. Now reasonably convinced the pulse-length was the problem, the team developed a software patch for the SOHO computer.

Two months after the initial failure the plan was put into action – and at the second attempt it worked.

When SOHO finally ceases operations, towards the end of the first decade of the new millennium, the helioseismology instruments on board, including the LOI, will have provided more than a full 11-year cycle's worth of high-quality data. There have been a few more glitches on the way. Some of the mechanisms in the GOLF instrument failed after the first month of uninterrupted observations. However, the GOLF team was able to work around the problem. The stability offered by the L1 point largely negated the impact of the loss of the mechanism and enabled the instrument to turn out cutting-edge science.

The solar physics community was, for a while, confronted with the possible loss of the entire mission. In the summer of 1998, communications with the SOHO spacecraft were lost during a period of planned manoeuvres. Contact was restored and observations resumed fully in November of that year. A gyroscope failure interrupted operations shortly after that, and in 2003 there were more communications problems to face. However, bearing in mind that in 1998 the mission was expented to last till 2003, the fact the SOHO scientists and engineers are confident it can continue to meet many of its original objectives possibly until 2008 bears testament not only to the spacecraft itself but also to the ingenuity of those who have provided work-around solutions to the problems.

8

FROM PARTICLE PHYSICS
TO COSMOLOGY

From the early observations and the first dedicated observatories, through to the roll-out of the major networks and the launch of SOHO, there have been impressive improvements to the quality of the helio-seismic data and an eye-catching diversification of the types available. Coupled with a steadily accumulating database, courtesy of the current long-term programmes, this flowering of observational helioseismology has, over the few decades of its history, led to dramatic advances in the science possible. Our aim in the next few chapters is to follow the history of this science, from the late 1970s through to the present day.

The discovery that not only the outer layers of the Sun but also the rest of its vast bulk pulsated had opened up the possibility of uncovering the structure, layer by layer, right down into the deep interior. The route to doing this would be via the use of inversion techniques, as trailed by Douglas Gough and Jørgen Christensen-Dalsgaard in their 'helio-logical' paper in 1976. By the early 1980s the p mode data were in place to enable in principle an inversion to be made. But first the detail of potential inversion strategies had to be fully worked out. Douglas Gough spent time developing these techniques and by about 1983 was in a position, together with Jørgen, to apply them to real data – initially, observations made by Jack Harvey, Tom Duvall and Ed Rhodes at Kitt Peak.

The first technique to be tested took the rotational frequency split-tings of the p modes as input and sought to find the rotational profile, as a function of depth throughout the interior, in those regions close to the equator. We will come to this in the next chapter. Here, we begin by con-sidering the first inversion made to recover an estimate of the sound speed as a function of radius. This was what the stellar modellers had been waiting for – a chance to see how well the best available models of the solar interior matched reality.

When they came, the results had far-reaching implications, for they led to revisions in our understanding of the physics of matter under the range of exotic temperatures encountered within stellar interiors. With the uncovered profile to hand it would also be possible to say something even more concrete about the ramifications of this internal structure for the solar neutrino problem.

Douglas based the first sound-speed inversion procedure on the import-ant finding of Tom Duvall that now carries his name: the Duvall law.

The first of Tom's eureka moments (he will reappear later with another) came one night while he was taking a break from analysing data at the Kitt Peak Observatory. While flicking through a newly published collection of reviews, called *The Sun as a Star*, he settled on an article by John Leibacher and Bob Stein about the 5-minute oscillations.[1] One part in particular grabbed his attention. John and Bob gave an equation that encapsulated the simple interference condition for waves making res-onant cavities in the solar interior. For us to fully appreciate the light this clicked on in Tom's mind it behoves us to think back once more to the analogy of waves in pipes.

Let us take as our simple model a pipe open at one end and closed at the other (recall that a severely truncated, sealed cone gives similar

---

[1] J.W. Leibacher and R.F. Stein, in *The Sun as a Star*, ed. S.D. Jordan (Washington: NASA, 1981), p. 263.

frequencies). From Chapter 3, we know the interference condition for standing modes in this semi-closed pipe demands the length be an odd integer number of quarter-wavelengths (save for a small end correction we shall again ignore here). Provided the simple dispersion relation holds (i.e. $\omega = c \times k$) this condition may then be re-written in terms of the ratio of wave speed (c) to frequency (ω), rather than in terms of wavelength ($2\pi/k$).

Next, we express the wave speed as the ratio of the size of the cavity and the time needed to traverse it. We then tidy up terms on both sides of the relation to arrive at a statement that says the following: the time taken for a wave to travel across the pipe is some integer multiple – *plus* one half – of the inverse of the frequency, ω, multiplied by a normalising factor equal to the mathematical constant π. We could have instead chosen a fully open pipe (recall that a cone complete to the tip gives similar frequencies). This would have given a similar relation; but the offset factor would have been unity, rather than one-half.

John and Bob showed how a similar relation could be written for waves trapped in cavities in the Sun. This follows because locally they can also be treated to a good approximation as plane waves. The travel time is now that across a given solar cavity, the frequency that of the oscillation mode formed and the integer the overtone number (the radial order, n) of the mode. The constant now depends on the types of boundaries that are most appropriate for describing the lower and upper turning points of the modes. We saw in Chapter 4 that these boundary conditions are not as simple as those of a mathematical pipe. However, the rest of the relation is the same.

When Tom read this section of the paper it did not take him long to realise it pointed to a way to collapse the ridges of the k–ω diagram onto a single curve. This curve would possess all the rich information contained in the numerous ridges, but all on one handy line. Instead of having to fit separate ridges this offered a way to simplify any analysis (both numerically and theoretically). Furthermore, an expression describing the curve would be a form of universal dispersion relation for the interior modes – an important addition to the canon of knowledge on the oscillations.

When a wave reaches the lower boundary of its cavity it must by definition be moving horizontally. The horizontal component of its speed – given simply by the ratio of its frequency and horizontal wavenumber – must then be the sound speed at this depth. No matter in which part of the k–ω diagram two modes are found, Tom recognised that provided they had this same ratio, ω/k, modes would penetrate to the same depth and be bounded by the same cavity. A common cavity implies a common travel time. So, by plotting the travel time – as expressed in terms of the observable quantities of mode frequency and overtone number – against the ratio of ω/k, he had found a way to represent the wealth of observational data in terms of one relation and a single curve.

This did not work right away. Tom had to vary the value of the trouble-some boundary constant to make the relation work. But when he had done so the procedure worked remarkably well. Tom had had his eureka moment (too late at night to tell anyone about it until the next day). The importance of this result was not lost on Tom or his colleagues. As he noted at the end of the paper reporting the finding, what was being plotted on this new form of diagram was the travel time of sound across each cavity, against the speed of sound at the bottom of each cavity. Surely this meant the plotted data could be 'directly inverted to obtain the speed of sound plotted against depth'.[2] This was the tack Douglas Gough initially took.

The first inversion using the Duvall law made use of an early set of Kitt Peak data and provided an estimate of the sound speed only in the outer 70,000 kilometres or so of the convection zone. But soon these data were supplemented by a much wider range of modes which enabled the sound speed to be estimated down into the radiative interior. Douglas recalls very clearly getting the result from this expanded set.

[2] T.L. Duvall, Jr, *Nature*, 300, 1982, p. 242.

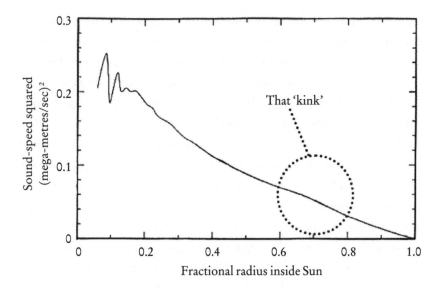

Figure 8.1. Square of the internal sound speed uncovered by Douglas Gough from his inversion of the data of Jack Harvey, Tom Duvall and Ed Rhodes. The kink marks out the base of the convection zone (see text). Adapted from figure in Christensen-Dalsgaard et al., *Nature*, 315, 1985, p. 378; reproduced with the kind permission of the Nature Publishing Group.

He was due to give a talk at a Royal Astronomical Society meeting in London the next day and was therefore working furiously to obtain a result he could announce. To the young scientist of today the idea of having to stay up well past midnight to get access to a mainframe computer may seem somewhat antiquated. This was often the only time of the day when sufficient computing time could be secured to run the numerically demanding code needed to solve astrophysical problems. The inversion of Jack, Tom and Ed's data placed such demands on both Douglas and the mainframe at Cambridge.

Douglas finally got his result in the small hours of the morning and set about plotting it to show the estimated sound speed as a function of radius inside the Sun (Figure 8.1). When Douglas turned the paper round and squinted to look down the curve edge-on, he could see a kink in the curve. It happened to be at a depth of roughly 200,000 kilometres beneath the surface of the Sun (a fractional depth of about thirty per cent). This was success number one – a telltale indication of where the bottom of the convection zone lay. As the first *direct* measure of the size of the zone this was an extremely important result, and confirmed the inferences made from the earlier helioseismic data.

By then only one other dedicated (possibly deadline-threatened) astrophysicist was present in the computer room. Keen to share the result, but mindful he wanted to be sure it was robust, Douglas asked his colleague to squint down the curve and mark the location of any kink. The field of expertise of the other astrophysicist lay well outside that in question (he was a cosmologist), but he placed his finger at the same location on the piece of paper. Elated, Douglas made a copy of the plot on a sheet of transparency and returned home to grab a couple of hours' sleep.

At the meeting he was therefore able to show that he and his colleagues had obtained the first estimate of the sound speed profile within the Sun. He pointed out the telltale kink in the profile and stressed the importance of knowing the depth of the convection zone in that it served as a useful diagnostic of the helium abundance. Because the curve was shown on a large screen, the audience were denied the opportunity to squint down the profile as Douglas and the cosmologist had done that morning. The audience therefore took a bit of convincing the kink was indeed present and Douglas recounted the story of the 'cosmologist in the computer room' to bolster his case. The kink was, though, undeniably present and the deep nature of the convection zone confirmed.

A comparison of the inferred magnitude of the sound speed with that of the best models would take a little more time to compute.[3] When this was done it was apparent immediately that the level of agreement between the two was very good, the difference being less than one per cent on average. But (and this was a big 'but') because the uncertainties on the

---

[3] Full details of the inversion were published in J. Christensen-Dalsgaard, T.L. Duvall, Jr, D.O. Gough, J.W. Harvey and E.J. Rhodes, Jr, *Nature*, 315, 1985, p. 378.

inferred profile were so small these differences were highly significant – several times the typical error bars – in the region between about thirty and seventy per cent of the solar radius. This lay in the radiative zone. Here, the sound speed in the seismically re-constructed profile exceeded that in the best model. One conclusion to be drawn was that the radiative zone was hotter than had been predicted.

A vital factor governing the temperature profile in the radiative interior is the opacity. The temperature gradient is directly proportional to the opacity in regions where radiation is the dominant method of energy transport. The inversion implied the efficiency of energy transport was lower than expected, beneath the convection zone. One way to bring the models into line would be to increase the opacity in the region just below the zone. This would have the effect of driving up the temperature in the radiative region in order to increase the temperature gradient and maintain the required luminosity.

Calculations suggested that a 10–20-per-cent rise in the opacity, appropriate to the few-million-degree temperature range just beneath the convective envelope, would give a significant improvement. The Los Alamos Laboratory in New Mexico had largely been responsible for calculating the opacities used by astronomers. An initial re-appraisal of their work suggested potential changes to the opacities would be insufficient to account for the sound-speed differences. It was then that the helioseismologists involved Forrest Rogers and Carlos Iglesias at the Lawrence Livermore National Laboratory in California. Forrest was interested in astrophysical problems and he and Carlos were persuaded to perform opacity calculations in their spare time for the temperature range in question.

With improvements made on the previous treatments, the suspicions of the helioseismologists were soon confirmed – the opacities of Forrest and Carlos were indeed higher than the previous values by about the right amount. With this result to hand, Forrest was able to secure formal support to perform a comprehensive series of astrophysical opacity calculations. These calculations would point out shortcomings in the previously used approaches and solve several long-standing astrophysical problems – all this from an inversion for the sound speed within the Sun.

The new calculations had a fairly modest effect on the opacity at a few million degrees centigrade. Only the unprecedented precision of the seismic data had allowed this discrepancy to be uncovered. At cooler temperatures, in the hundred-thousand-degree range, the effect was much larger. These temperatures are encountered in the middle of the Sun's convection zone. Here, the structure is unrelated to the radiative opacity and so this enormous effect would never be apparent in the solar data. However, it did matter for other types of star. The first long-standing puzzle to be solved was the Cepheid mass problem.

We have seen that Cepheid variables are classical pulsators, which rely on favourable changes in opacity during cycles of compression and rarefaction to drive their oscillations. The region of driving within these stars happens to lie in the few-hundred-thousand-degree temperature range. It is important to recognise that the Cepheid story had important input independent of helioseismology. As early as 1982, Norman Simon had written a paper suggesting inaccuracies in the opacities used by astronomers might be responsible for the problem. Two Danish astronomers, Andreasen and Petersen, predicted the required change, to a very good level of accuracy, in 1988.

The ratio of the pulsation periods of those Cepheid variables observed to oscillate in more than one mode is a useful indicator of the stars' mass. The value the ratio takes depends upon the pair of modes being observed – they could be the fundamental and first overtone, or perhaps the first and second overtones – and the distribution of density inside the star. The 'mass problem' arose because masses determined from pulsation theory – using the observed period ratios – failed to match those predicted from stellar evolution theory, the second approach needing to evolve the star to a point where it could in principle pulsate. Estimates from the pulsation technique were found to be on average about fifty per cent smaller than those from evolution theory.

Other mass determination techniques, which made use of oscillation data, also gave values that were too small. For example, the 'bump-mass' method relies on matching the observed luminosity-versus-time curve of the pulsation with one from theory. Cepheids with oscillation periods of around 10 days exhibit bumps on their curves. The bump arises from a resonance between the fundamental and second overtone. Its position on the curve depends upon the period and is therefore another sensitive measure of mass. Pulsation calculations were unable to match the observed curves when the masses used as input to the computations were chosen to be consistent with those from stellar evolution theory.

However, the new opacities solved the bump-mass and period-ratio problems in one go. The effect on the pulsation calculations was sufficient to increase the resulting mass estimates by about the required amount. The new opacities were to perform a similar service for the pulsationally determined masses of RR Lyrae stars. These classical pulsators are extremely old and have a very low heavy element content. We recall that metals influence the opacity very strongly. That the new calculations increased the opacities of these stars, by enough to solve their mass problem, was therefore a surprise.

The new data also resolved a more perplexing question: why did β Cephei stars pulsate at all? These are very hot, massive stars that reside outside the range of classical pulsators. Nevertheless they exhibit strong oscillations. Wojtek Dziembowski had worked for years to understand why they pulsated. An opacity-driven mechanism seemed to be inadequate to give the observed modes. However, with the new opacities the mechanism now worked.

These spin-off results were an important success for helioseismology. One should also not lose sight of another implication, somewhat closer to home – what the inferred structure meant for the solar neutrino problem. For those who felt that misunderstandings of conditions in the deep solar interior were not the cause, the news was good. In spite of the adjustments to the opacity which the helioseismic observations had demanded, the agreement with the models was such that an anomalously cool interior continued to appear unlikely. Nevertheless, it should be noted that this first inversion had only just penetrated the energy-generating core, and this was where the uncertainties in the determination of the sound speed were largest. More data, from the very lowest-degree modes, would therefore be needed to get a clearer picture of the core.

It is perhaps worth discussing at this stage what inversions of p modes can achieve in different parts of the Sun.

First, the precision with which one can infer the structure of the core is always going to be inferior to that achievable in the layers above. Nature ensures this is so. Not only do the waves that form the acoustic modes spend less of their time sampling deeper, as opposed to shallower, layers – because the sound speed increases with depth – but there are relatively few modes that penetrate the core and so there is less information available than for the outer parts of the interior. The low-degree modes are also split by the internal rotation into fewer components than their higher-degree counterparts. The frequencies of the core-penetrating modes are therefore less well determined than their shallower cousins'. This affects inversions for both the static (the sound speed or density) and the dynamic (rotation) structure of the deep interior.

Thanks to the richer splittings of the medium- and high-degree modes it is possible to invert for structure in both depth and latitude in the outer layers. The outer components in a multiplet have the largest splittings and probe regions closest to the solar equator; as one moves in frequency towards the centre of the multiplet one encounters components that sample successively a greater fraction of the higher latitude regions. This differential penetration property in latitude is what enables an inversion to be in principle performed in latitude. Just as with an inversion in depth, you need a wide range and large number of modes to obtain a respectable resolution. Because the low-degree modes are split into a small number of components it is not possible to make a meaningful inversion in latitude in the deeper parts of the radiative interior and core.

Observing for long periods, and detecting cleanly as many of the low-degree modes as possible, is important for refining inferences about the core's properties. Refer back to our discussion, in Chapter 7, of the motivating factors that led to the establishment of the long-term observing programmes. Longer datasets mean the frequencies, splittings and other parameters of those modes already present can be measured to even higher precision. What is more, new low-frequency modes can emerge from the background noise. The rich benefits to be had from detecting these long-lifetime, low-frequency modes have already been made clear, specifically that their frequencies and splittings can be determined to exquisite precision and accuracy. The more precise and accurate are our observations, the more precise, reliable and demanding will be the science we can extract from them.

The goal of uncovering very low frequency modes is in many respects the Holy Grail of observational helioseismology. This means not only working our way down to the low-overtone p modes – and ultimately the fundamental breathing mode – but also unearthing the gravity, or g, modes.

Interior gravity waves are expected to be found in cavities beneath the convective envelope (see Chapter 3). This is both a good and bad thing to an interested helioseismologist. First let us consider the good part. Unlike the sound waves, which spend most of their time in the cooler, outer parts of the Sun, the buoyancy waves are forced to spend all of their time in the radiative interior. Their properties are therefore determined completely by the conditions there. For a scientist interested in uncovering the secrets of the Sun's core this is splendid news. A given change in the structure of the core will give rise to a much larger fractional change in the frequency of a g mode than of a p mode.

So, the gravity modes are a much more sensitive probe of the core than their acoustic cousins. But what about the bad part? This is a little bit more complicated. Remember the Sun is opaque to radiation. When you observe the Sun what you see is light from its visible surface, the photosphere. It is also possible to see the emission of light from higher up in the atmosphere. But one cannot see the layers beneath. This has implications for how the oscillations are observed. We are forced to detect the changes at locations in the photosphere, or above. This is a big problem if you want to observe gravity modes trapped beneath the convection zone. Although it is possible for them to nudge the layers above up and down, thereby allowing their signature to be observed, in theory, in the photosphere (via the evanescent waves), the higher up one goes above the base of the convection zone the smaller the effect becomes. Alas, calculations predict that by the time anything left over reaches the photosphere its amplitude is so small as to make it barely detectable. But this is not about to stop determined observers. The scientific rewards to be reaped by a positive detection are just too great to ignore.

To put the extent of the challenge in context, consider the following. The strongest whole-Sun p modes give rise to a signal at the photosphere that moves the surface at a velocity of several centimetres per second. Over a complete 5-minute cycle of oscillation this displaces the surface by about 50 metres – not very much for a body 700,000 kilometres in radius. Nevertheless, to the finely honed helioseismic instruments currently trained on the Sun, this represents a huge signal. But what are the strengths of the strongest g modes?

At the time of writing I think it is safe to say that no one would claim to have actually observed one of these modes. This has not been for want of trying. We must therefore rely instead upon a theoretical calculation of the amplitudes they might give rise to. Performing calculations of this nature for p modes is far from trivial. Doing so for the as yet unobserved g modes is even less straightforward. This is because it is far from clear what mechanism, or mechanisms, might dominate their excitation. If we take the current best, and most optimistic, estimates available these point to Doppler velocity amplitudes of maybe a millimetre per second for a g mode with a period of about 40 minutes. This is about a factor of twenty, in amplitude, lower than for the p modes. Most analysis is done in terms of power – the difference is then a factor of four hundred. Put another way, a back-of-the-envelope calculation suggests a maximum surface displacement of just over 2 metres. See the problem? But that is not all. Matters are made worse still by the fact that background noise from the convective granulation increases in strength at lower frequencies; furthermore, achieving excellent instrumental stability over long periods can pose many problems for the observer. The challenge has been laid down, and many helioseismologists are more than willing to take up the cudgel. I shall leave a brief discussion of where we stand with regard to this search to the conclusion of the book.

Throughout the early 1980s searches for core-penetrating g modes continued, with even a few claims of detections. But these claims were controversial and effort remained focused on exploiting the well-established p-mode data to their fullest potential.

Several studies began to highlight the use of the fine spacings of the low-degree modes – the separations in frequency between adjacent $l = 0$ and $l = 2$ or $l = 1$ and $l = 3$ modes – as a means of discriminating theoretical models of the Sun with extra features included that, if valid, could provide a solar solution to the solar neutrino problem. The magnitudes of the fine spacings are a fairly robust and sensitive probe of conditions in the Sun's deep interior. They are robust because they are a difference between two very similar mode frequencies. Since, to a first approximation, the way each mode in a pair samples the outermost layers is similar, taking the difference largely removes the influence of these complicated layers.[4] Model calculations are therefore rendered more reliable because the poor modelling of the surface has little influence on the predicted values.

The mode properties may be similar in the outer layers, but they are most certainly not in the deep interior. The higher-frequency mode in each pair is always formed by waves that penetrate deeper than those of

---

[4] Recently, Ian Roxburgh and Sergei Vorontsov have proposed the use of the ratios of the fine to large (overtone) frequency spacings as probes of the deep interior (with an eye on their potential use for seismic investigations of other Sun-like stars). These ratios are more immune to the influence of the surface layers than are the fine spacings.

its lower-frequency counterpart. The difference in frequency is there-fore sensitive to conditions in the deeper-lying layer the higher-frequency mode samples but the other mode does not.

A more careful look at how modes sample the interior volume reveals that conditions in the layers outside the core contribute as much to the value of the spacing as do those in the core itself. However, if the struc-ture of the Sun is altered (by adjusting various model parameters) the contribution from the layers outside the core hardly changes at all, but the contribution to the spacing from within does – by an amount which can be measured.

The use of the spacings to test models with features that might solve the solar neutrino problem first found real voice in two independent studies, one by John Faulkner, Douglas Gough and Mayank Vahia, the other by Werner Däppen, Ron Gilliland and Jørgen Christensen-Dalsgaard. Both appeared in the same edition of *Nature*, in 1986, and looked in detail at models with 'weakly interacting massive particles', or WIMPs, in their cores. (Yes, there are MACHOs in the lexicon of astro-physicists as well.)

Cosmologists introduced WIMPs as a means of solving the so-called missing mass problem. Observations of distant supernovae – remarkably bright, intense releases of energy marking the death throes of massive stars – have allowed cosmologists to probe the history of the expansion of the universe. These and other data appear to have established that most of the universe comprises material distinct from the matter of which we are made. About a third is believed to be in the form of dark matter – matter that cannot normally be seen and which is therefore apparently 'missing'. Cosmologists like to talk in terms of two classes of this dark, missing matter: hot dark matter and cold dark matter; and WIMPs can come in both types.

As their name implies, WIMPs interact very weakly with normal matter, which makes them hard to detect experimentally. Hot dark matter WIMPs travel very fast – they move at velocities typically a good fraction of the speed of light. In contrast, cold dark-matter WIMPs dawdle around at much slower speeds. Neutrinos are a hot dark matter WIMP candidate. But to be so they must have mass, an important point we shall return to as our story of the solar neutrino problem unfolds. However, cold dark matter scenarios are more favoured by cosmologists today. This is because the hugely complicated simulations cosmologists run to model the evolution of structure in the universe work much better – in the sense of giving a distribution of matter which looks more like the current universe – if cold, rather than hot, scenarios are adopted.

MACHOs on the other hand are cold dark matter candidates, 'massive compact halo objects', residing in halos surrounding galaxies. They are not postulated to be individual sub-atomic particles, like WIMPs, but large structures like brown dwarfs (failed stars not big enough to shine) and planets.

So, how might WIMPs fit into a solar model? WIMPs can be captured gravitationally if they collide with solar matter. Those unfortunate particles that become trapped within the confines of the solar core can have an interesting effect on the conditions. This is because WIMPs act as efficient transporters of energy.

The dominant energy transport mechanism in the solar interior is radiation. Each photon that carries some of this energy can travel only a fraction of a centimetre before it collides with another particle. In contrast WIMPs encounter no such difficulties because they are 'weakly interacting'. The typical distance they can travel between collisions is postulated to be of the order of the solar radius itself. This means a WIMP can pick up some energy in the core and carry this energy all the way to the outer layers of the star before interacting again and dumping the energy. Since WIMPs therefore provide an additional means of transporting energy outwards the internal temperature gradient, and therefore the central temperature, of the Sun need not be so high in order to give the observed luminosity.

The size of effect WIMPs have on the structure of the deep interior depends on their intrinsic properties. For example, heavy WIMPs are sluggish. They possess characteristic velocities that are smaller than those of their lighter counterparts. Heavy WIMPs are therefore constrained in a small volume in the core, whereas lighter WIMPs can in principle occupy a larger volume, giving them the potential to influence the structure of a bigger fraction of the core. A proportionately larger affect on the structure is also given by increasing the number of WIMPs that can be captured, or by raising the likelihood of their interacting with normal matter (a scientist will then talk in terms of increasing the 'interaction cross-section').

In the late 1970s, John Faulkner and Ron Gilliland made the first calculations of the effect WIMP accretion would have on the Sun. Their interest in the problem was rekindled, in the mid 1980s, by the work of David Spergel and William Press, who 'explored the same general idea from a fresh perspective'.[5] The helioseismic studies of Faulkner et al. and Däppen et al. were to follow soon afterwards.

Scientists must first hypothesise a set of physical characteristics for WIMPs. Once they have been created on paper in this manner a solar model can be built that contains them, and the resonant properties of the model can be estimated for comparison with the observed mode frequencies. When Faulkner et al. and Däppen et al. performed such analyses they reached the same conclusion. The predicted fine spacings of models having WIMPs with masses about five times that of the proton seemed a better match to the early helioseismic observations than did those of models that were WIMP free. However, this initial, intriguing assessment was destined not to stand the test of time. Improvements in the precision of the observed frequencies (and by implication the fine spacings), coupled with adjustments to the way the models and model frequencies were calculated, soon brought the observations and the standard, WIMP-free, models into line. This was first demonstrated by three studies presented at the hugely successful Tenerife helioseismology conference in September 1988. Bernard Gelly and his colleagues from the Nice group and Art Cox, Joyce Guzik and Russell Kidman both

---

[5] J. Faulkner and R.L. Gilliland, *Astrophysical Journal*, 299, 1985, p. 994.

compared model fine spacings with those from a selection of observations. Douglas Gough and Sasha Kosovichev performed structure inversions to show the aforementioned WIMP models could indeed be excluded (by comparison of the observed and modelled sound-speed profiles). Come 1990 the Birmingham group, now operating a three-site network, was able to use its extended data to rule out these same WIMP models at a high level of significance.

The intervening decade and a half has seen further substantial improvements in the quality of the helioseismic fine spacings and the sound-speed inversions. It is now possible to place even tighter constraints on the possible presence of WIMPs in the solar core. In the case of WIMP mass, this in effect means raising the allowed lower limit to the mass. In short, the heavier the WIMP, the less effective a job it does of modifying the internal structure. Since the helioseismic observations appear to be such a good match to the standard solar model, only those models with heavy WIMPs can be admitted as possible.

A recent study, in which helioseismologist Ilídio Lopez collaborated with cosmologists Joe Silk and Steen Hansen, has set a new lower limit of about sixty times the mass of the proton. In conclusion, it is now clear that if WIMPs are present they are unlikely to have what it takes to play an important part in the solar neutrino problem. It goes without saying that uncovering the signature of their presence in the solar core would be of huge significance for cosmology.

The late 1980s and early 1990s heralded the inclusion in the solar model of non-standard processes – 'non-standard' in the sense of not usually being included in the models – that have stood the test of time. However, this spelt bad news for those looking for a solar solution to the solar neutrino problem. The effects, heavy element settling and diffusion, turned out not to have a significant impact on predicted neutrino fluxes. But they did give rise to important adjustments to the picture of the Sun's internal structure and interesting consequences for cosmology.

Settling causes helium and heavier elements to settle deeper into the interior at the expense of the lighter hydrogen they displace. The effect arises from the momentum exchanged when heavy particles collide with lighter ones. This transfers momentum to the lighter elements. The typical velocities at which the heavier elements move around are reduced; these lower 'thermal' velocities result in the particles being confined more closely to the centre of the star by the Sun's internal gravity. The tendency is, therefore, for the heavy elements to sink as time passes, hence the term 'gravitational settling'. This settling is, however, inhibited by the second effect – diffusion. Diffusion is a product of random collisions of particles with other particles, or particles with photons. It results in particle or thermal concentrations becoming more diffuse (as the name suggests).

The notion that diffusion might take place in stars was nothing new. As early as 1917, Sydney Chapman (and in 1926 Eddington) had reached the conclusion that the effects were so small as to be of little importance. What was different in the late 1980s was an appreciation that the fine probe of helioseismology might make the effects observable and testable. (Noerdlinger was the first scientist to study in detail the effects of helium diffusion in the Sun in the late 1970s.)

Settling and diffusion leave their mark beneath the convection zone, because within it material is always well mixed. A subtle but steady rearrangement of the distribution of the constituent elements that make up the radiative interior can result from these effects; this represents an important modification to the internal structure of the star. Adjustments also take place on a very long timescale; indeed, the diffusion timescale is so great that equilibrium between the settling and diffusion will never be established during the lifetime of the Sun.

The late 1980s and early 1990s brought a renewed interest in diffusion and settling, several studies taking a fresh look at the implications for the Sun.[6] Gravitational settling and diffusion of helium and heavy elements

---

[6] Those involved included the likes of John Bahcall, Jørgen Christensen-Dalsgaard, Art Cox, David Guenther, Joyce Guzik, Russell Kidman, Abraham Loeb, Georges Michaud, Marc Pinsonneault, Charles Proffitt and Mike Thompson.

were introduced into the up-to-date models and it was found this changed the speed of sound beneath the convection zone by, on average, about 0.5 per cent. To solar modellers in the 1970s this would have been viewed as a size of effect no one was in a position to test. Over a decade later the solar oscillation data were good enough to uncover the difference. Papers by Jørgen Christensen-Dalsgaard, Mike Thompson (a former student of Douglas Gough) and Charles Proffitt and by Art Cox and Joyce Guzik demonstrated that predictions from solar models gave a significantly better match to the observed helioseismic data when settling and diffusion effects were included. Another real success.

Scientists in the field were now beginning to reveal in some detail the structure of the solar interior. The outcome of the settling and diffusion investigations also fell into a trend towards seismology results that had implications for other areas of astrophysics. This time the spin-off came in how stellar evolution theory was used to calibrate the ages of the oldest stars in the universe, which serve as a lower bound on the age of the universe itself.

Globular clusters each contain about a hundred thousand or so ancient stars. We know these stars must be very old because the spectrum of light they emit betrays little evidence for heavy metallic elements, which pepper the solar spectrum with Fraunhofer lines. Remember that stars are the factories that forge these heavy elements by successive rounds of nuclear fusion. The evidence therefore indicates that globular clusters were formed at an early epoch of the universe.

To arrive at an estimate for the age of a cluster[7] demands two things: observations of the intensities and colours of the stars that make up the cluster, and a good model to describe how the stars have evolved over time. Helioseismology gives direct input to the second of these. To understand how, we must first look at the information the other observations provide.

[7] An overview of the determination of ages of globular clusters can be found in B. Chaboyer, *Physics Reports*, 307, 1998, p. 23.

An intensity–colour plot – or Hertzsprung–Russell diagram – can reveal much about a cluster. From this simple diagram it is possible to uncover the common life history of the stars and to characterise the different types of star present. When we look at the night sky with the naked eye or a small telescope, we can see stars of different brightness and, occasionally, stars showing a hint of a red or blue hue. The colour tells us about the surface temperature of the star. Stars that are blue in appearance are hotter than their red compatriots, meaning the peak of their emitted energy is at the shorter-wavelength end of the visible spectrum.

Brightness tells us how luminous a star is. This depends on three things. Two are intrinsic properties of the star – the surface temperature and radius. The larger both are, the brighter the star appears. The third factor is the distance the star lies from us. We can remove the dependence of the intensity–colour plot on distance – so that what remains is fixed by aspects of stellar evolution only – by choosing to look at a collection of stars that are all about the same distance away, like those in a cluster. Furthermore, since the stars in a cluster all formed at about the same time – from a large, collapsing cloud of gas which fragmented into the smaller parts from which individual stars formed – our plot will contain factual information for objects that possess a common origin. This is why the plot reveals to us the life history of the cluster.

The plot has an ordered appearance (Figure 8.2). Stars on the Main Sequence, which are, like the Sun, burning hydrogen into helium in their cores, occupy a diagonal strip on the plot. The position each star occupies alters only a little during the Main Sequence lifetime. Stars at the top end of the Main Sequence have higher surface temperatures, are intrinsically brighter and are also more massive than those on the lower Main Sequence. Even though these heavier stars have more fuel to burn, their high core temperatures mean they use it up quickly. Whereas the Sun is set to take several thousands of millions of years to burn its core hydrogen, a star a hundred times more massive will do so in a few thousand years.

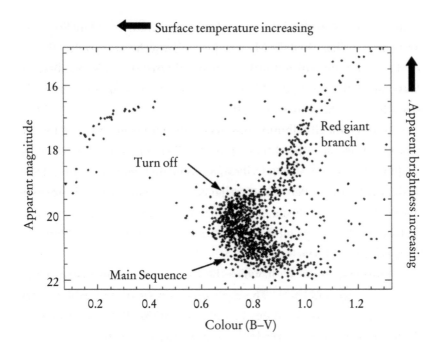

Figure 8.2. A colour–magnitude diagram for just over a thousand stars in a globular cluster (NGC 5286 in the constellation Centaurus). After stars leave the Main Sequence they ascend the red-giant branch. The colour index is a measure of the difference in the brightness of each star in the blue (B) and visual (V) (yellow) wavelength parts of the spectrum. Data from N. Samus, A. Ipatov, O. Smirnov, V. Kravstov, G. Alcaíno, W. Liller and F. Alvarado, *Astronomy and Astrophysics Supplement Series*, 112, 1995, p. 439; and provided by the CDS VizieR service.

Once its core supply of hydrogen is exhausted – because it has all been burned to helium ash – a star is faced with an energy crisis. The innermost parts of the star collapse. Energy is made available from this contraction and eventually temperatures rise sufficiently to allow an annulus of hydrogen to be ignited surrounding the core. The outer layers of the star also adjust to the changes deeper down and expand, and here the star peels away at right angles from the Main Sequence – it is no longer a Main Sequence star.

The first stars to move off the Main Sequence are the most massive. As time winds on, successively lighter stars follow suit, meaning a 'banana skin' of evolved stars peels back, shortening the Main Sequence as it goes. The length of the Main Sequence must therefore reflect the age of the cluster. It turns out that a robust indicator of age is the observed location of the tip of the Main Sequence on the brightness axis, the so-called Main Sequence turn-off.

Although the distribution of stars on the plot is not affected by distance, the absolute brightness scale is. The turn-off point must be calibrated to absolute units so it can be compared with predictions from stellar evolution models. These stellar codes provide the sought-for estimate of how old the cluster must be to have the observed turn-off. So, the distance to the cluster must be estimated; and a good theory of how the stars evolve must be to hand.

There are several ways in which the distance may be estimated, which we will not go into here. The age estimate rests on the evolutionary models. Once helioseismology had shown that the effects of settling and diffusion were at work inside the Sun, it was only natural to suppose they would be present in other stars too. The appropriate inclusion of these effects in the models changed the time they took to evolve to the observed cluster turn-offs. The cluster ages were, as a result, revised downwards by about seven per cent to current typical values of about 12,000 million years.

With a little thought (and an admittedly simplified argument) we can understand why settling pushes the age estimate this way. What settling does is to increase the amount of helium in the deep interior by a means other than the evolutionary burning of hydrogen into helium. If you are not aware the effect exists you will be fooled into thinking the observed concentration arises solely from the effects of nuclear fusion. So models without settling need to be evolved to ages older than their settling-laden counterparts to match what is there. Things are actually not quite as simple – evolution and settling leave different characteristic imprints throughout the radiative interior and core. But you get the idea.

The inclusion of settling and diffusion was not the only helioseismic spin-off for the cluster-age story. Another seven per cent was shaved off the ages as a result of the inclusion of a better equation of state in the models. Helioseismology can test the physical state of matter in the solar interior to a high level of precision. The convection zone is particularly well suited as an astrophysical laboratory in which to do this. Its structure is, crucially, independent of the radiative opacities and all the complications involved in determining reliable values for them. This makes it much more straightforward to diagnose the impact on the structure of tweaks made to the equation of state.[8] It has been shown – thanks to work by several leading players, including the enthusiastic Werner Däppen – that slight modifications to the ideal description mentioned in Chapter 2 are needed. Diagnoses of this type would not have been possible without the ultra-fine probe offered by the modes.

By the early to mid 1990s it was becoming increasingly clear that the standard solar model really was a good match to the Sun, at least at a level that mattered for the solar neutrino problem. Non-standard models incorporating low heavy element and helium abundances, and also those with low-mass WIMPs, had all suffered rough treatment at the hands of helioseismology. Settling and diffusion were important, non-standard processes now being incorporated into the models, but their effects had only a modest impact on neutrino production.

At the same time more data were being collected on the neutrinos themselves. Ray Davis's chlorine experiment had continued operation, and observations from other instruments were now augmenting his precious data. First, Japanese and American physicists adapted an experiment set up in a mine in the Japanese Alps to allow it to detect very high-energy solar neutrinos scattered off electrons in water. The Kamiokande experiment was replaced in the mid 1990s by an even bigger version, appropriately named Super-Kamiokande. Two further experiments came on line in the 1990s. The Russian–American Gallium

---

[8] More formally: the structure depends principally on the equation of state, the composition and the specific entropy. In the thin layer just beneath the solar surface, matters become more complicated.

Experiment (SAGE) and European Gallium Experiment (GALLEX) detected low-energy neutrinos generated by part of the solar reaction chain that converts two hydrogen nuclei into a deuterium nucleus; neutrinos not accessible to Ray Davis's detector.

Since the new experiments all uncovered a significant deficit of neutrinos compared with the numbers expected, the reality of the solar neutrino problem was further reinforced. Although result after result from helioseismology pointed strongly to the need for a solution in the physics of the neutrinos, it took time for the particle physics community to take these ideas on board. This is not surprising: helioseismology, although going through a period of rapid expansion and success, was a very new field, with which even many astronomers and astrophysicists were less than familiar. John Bahcall, who since the early 1960s had focused his efforts on getting to the bottom of the solar neutrino problem (see end of Chapter 2), felt the tide began to turn in the later 1990s. John had begun to think seriously about making use of helioseismology in the mid 1980s. This led to the publication, in 1988, of the results of a systematic study of the diagnostic potential of oscillation data, made in collaboration with Roger Ulrich. John's interest in applying helioseismic data was piqued in 1995 by the publication of the first results from Steve Tomczyk's LOWL instrument, housed permanently at Observatorio del Teide on Tenerife. LOWL had a very important contribution to make during this stage of the history of the field because it was able to observe modes covering a wide range of angular degrees for very long periods. The GONG network, and some of the instruments on the SOHO spacecraft, would soon provide similar high-quality data. However, the various GONG sites were then only just starting to come on line, and SOHO would not commence regular operations until early 1996.

With LOWL, it was possible to probe a sizeable fraction of the solar interior, down into fairly deep-lying layers. The data of course had limitations. They were collected from only a single site, which meant there were substantial gaps in coverage each day. However, the problems caused by the gaps were outweighed at the time by the advantages that accrued from having all the needed modes in a single dataset. The subtleties of making a set of frequencies for analysis from combinations of observations of different instruments were then poorly understood. An all-in-one set was therefore a very handy thing to have.

John contacted Steve in the summer of 1996 to ask whether he had estimates of the sound-speed profile available from the LOWL data. Steve referred John to Sarbani Basu and Jørgen Christensen-Dalsgaard, who were working on the observations in Denmark and subsequently sent John the results of Sarbani's analysis. When he plotted the difference between the helioseismic and standard-model sound speeds, John was 'astonished by the precise agreement' between the two. In his words, 'it seemed too good to be true'.

It is important to recognise that the oscillations do not directly probe the temperature (one of the key parameters in fixing the rate of neutrino production). Rather, they provide a means of re-constructing the sound speed, which depends upon both temperature and composition. Inference about the temperature profile depends, therefore, on the assumed trend of composition throughout the Sun. It is possible to arrive at a profile for which temperatures are low enough to give the observed neutrino fluxes in the terrestrial detectors. However, the nature of the profile is then so contrived as to be completely 'unphysical'. In fact, several studies have shown conclusively that by pushing the composition profile to the boundary of being unphysical – so that it still gives a stable Sun with the observed radius and luminosity – it is not possible to reduce the central temperature of the models by enough to resolve the solar neutrino problem.

The inferred temperature profile does not quite tell the whole story. In order to get a proper prediction for the neutrino flux, hard numbers are needed for the rates at which the various nuclear reactions occur. The rates depend upon the sensitivity of different parts of the nuclear reaction chain to the conditions in the core and the cross-sections offered by the particles involved in the reactions, i.e. how big each particle appears to its fellow reactants and therefore how likely they are to interact.

In the central energy-generating core the agreement between the helioseismic and standard solar-model sound speeds was better than 0.1 per cent. This was far too small a difference to admit the possibility of a Sun with a low central temperature. Realising this was a powerful argument in favour of needing a particle physics solution, John quickly wrote a paper for *Physics Review Letters*, on which Marc Pinnsoneault, Jørgen and Sarbani were co-authors.[9] John felt this paper went a long way in helping to convince some physicists of the need to look beyond solar physics for the solution. These conclusions were reinforced when, a year later, John, Sarbani and Marc published a study that looked at just how uncertain the solar neutrino flux predictions really were.

By the turn of the millennium the helioseismology community was able to make its case even more forcefully. The high-quality, core-penetrating datasets were now even longer, allowing more precise conclusions to be drawn about the sound speed in the solar core. In studies made with data collected by the GOLF and MDI instruments, on board the SOHO spacecraft, Sylvaine Turck-Chièze and her colleagues used the tiny differences uncovered between the helioseismic and model sound speeds to tweak the solar model to minimise the differences in the radiative interior.[10] The helioseismically tuned solar models given by the analysis produced fluxes still significantly below those of the neutrino observations.

And then, in 1998 and 2001, new results from the neutrino detectors combined to confirm what helioseismology had been indicating for over a decade – the problem really was in particle physics.

In the late 1960s, two Russian theoreticians – Bruno Pontecorvo and Vladimir Gribov – had proposed that neutrinos might be able to switch states – or 'oscillate', to give the effect its formal name – changing in a

---

[9] J.N. Bahcall, M.H. Pinsonneault, S. Basu and J. Christensen-Dalsgaard, *Physical Review Letters*, 78, 1997, p. 171.
[10] S. Turck-Chiéze et al., *Astrophysical Journal*, 555, 2001, p. L69; see also S. Couvidat, S. Turck-Chiéze and A.G. Kosovichev, *Astrophysical Journal*, 599, 2003, p. 1434.

chameleon-like manner from one flavour to another. Since Ray Davis's chlorine detector, and the SAGE and GALLEX detectors, could only see the electron flavour, an effect of this type would offer a way out of the solar neutrino problem. What if, during their journey to the detectors on Earth, some fraction of the electron neutrinos made in the core changed flavour? This would kid the scientists into thinking an insufficient number of neutrinos were being made when in fact there were enough being made *but* a number were going missing by turning into the undetectable muon and tau flavours.

The 1998 result, announced by the Super-Kamiokande collaboration, did not actually involve solar neutrinos, but it provided the first direct evidence for the neutrino oscillations.[11] The Japanese detector observed neutrinos made in the Earth's atmosphere from collisions of high-energy cosmic rays with oxygen and nitrogen. Whereas nuclear reactions in the solar core make electron neutrinos, this atmospheric reaction gives rise predominantly to muon neutrinos.

Unlike the other neutrino detectors, Super-Kamiokande was also sensitive to the muon flavour (in addition to the electron flavour). What is more, Super-Kamiokande had directional sensitivity. By tracking paths followed by particles after scattering it was possible to re-construct the direction from which incoming neutrinos had come. This additional capability had allowed the original Kamiokande detector to show that electron neutrinos observed by the terrestrial detectors had indeed originated in the core of the Sun. Now, Super-Kamiokande was used to look for differences in fluxes of neutrinos, made in the Earth's atmosphere, which originated from opposite directions – high above the detector in the Earth's atmosphere, on the one hand, and from the atmos-phere on the opposite side of the planet, on the other. Because of their weak interaction with normal matter, the second cohort of neutrinos was able to make the journey through Earth to the detector. Scientists found

---

[11] The Super-Kamiokande Collaboration, Y. Fukuda et al., *Physical Review Letters*, 81, 1998, p. 1562.

that roughly half of the muon neutrinos taking this longer path went missing, whereas their compatriots following the much shorter path from above did not show the deficit. The implication was that some of the muon neutrinos were revealing their chameleon tendencies and turning into undetectable tau neutrinos. Neutrino oscillations were a reality.

Although this dramatic result opened the way to a possible particle physics solution to the solar neutrino problem, it was not enough to make a cast-iron case. Evidence for oscillations in the solar neutrinos themselves was still lacking. The data compiled by the atmospheric observations were relevant to muon and tau, not electron, neutrinos. Furthermore, the phenomenon had been observed under a different energy regime – the atmospheric neutrinos had energies about one thousand times higher than their solar cousins'.

The second of the aforementioned, and crucial, neutrino detector results (from 2001) went a long way to removing some of these potential question marks. It arose from data added by a new detector, the Canadian-based Sudbury Neutrino Observatory. This observatory was designed to run in different modes of operation. This capability would ultimately make it sensitive to all three different flavours of neutrinos. In the first mode, it collected data on the electron neutrinos, the only flavour made by the nuclear reactions in the solar core. From all the generated electron neutrinos it would only be able to detect those with very high energies – the low-energy threshold of the detector was such that it simply could not detect neutrinos produced by other parts of the reaction chains.

The particles it could detect also happened to be the very same neutrinos to which Super-Kamiokande was sensitive. However, there was a crucial difference. Whereas the Sudbury Neutrino Observatory only sensed the electron flavour, the Japanese observatory could also detect the signature of muon neutrinos in the same energy regime. With the chameleon-like oscillations in operation, the physicists would have expected the measured flux from the Japanese detector to be higher than that from its Canadian counterpart, since Super-Kamiokande offered a way of picking up missing electron neutrinos that were now muon in flavour. The results, when they were announced in 2001, bore out this rationalisation of the discrepancy.[12] The conclusion was reinforced when the Sudbury Neutrino Observatory entered a new mode of operation in which it was also able to detect the other flavours.[13]

The picture that has emerged is therefore one in which the neutrinos are able to change flavours. Because the electron neutrinos can oscillate into one of the other neutrino states, their flux at the Earth is reduced. It is this new property of the neutrinos that therefore appears to resolve the solar neutrino problem – not an adjustment to the physics of the Sun's interior.

Scientists are now seeking to understand flavour change and what drives it. The essence of Pontecorvo and Gribov's original idea was that the solar electron neutrinos could switch states while in transit between the Sun and Earth. The longer the journey, the greater would be the chance of a change in flavour taking place. The terrestrial atmosphere observations demonstrated very clearly that distance matters in this process. However, another type of oscillation has also been proposed in which the chameleon tendency is enhanced when the neutrinos pass through matter. Lincoln Wolfenstein, and independently Stanislav Mikheyev and Alexei Smirnov, showed that the electron neutrinos might be able to interact with electrons in the solar interior, thereby giving a type of resonance that could drive changes in flavour. These three physicists now lend the initials of their surnames to what is called the MSW effect.

Regardless of which effect dominates, the fact that neutrinos can change flavour presents a major challenge for fundamental physics. I have referred often to the standard model of the solar interior. Physicists of all types

[12] SNO Collaboration, Q.R. Ahmad et al., *Physical Review Letters*, 87, 2001, id. 071301.
[13] SNO Collaboration, Q.R. Ahmad et al., *Physical Review Letters*, 89, 2002, id. 011301; SNO Collaboration, S.N. Ahmed et al., *Physical Review Letters*, 92, 2004, id. 181301.

(or flavours) like the reassuring tenor of these words. It should therefore come as no surprise that fundamental physics has its standard model too. However, this is The Standard Model. It seeks to describe the behaviour of matter at the sub-atomic level – the very nature of the particles that inhabit the universe – and the forces through which matter interacts.

The Standard Model incorporated within it a means by which particles could acquire their mass. The mechanism depends on a special (as yet undetected) particle called the Higgs boson. The idea is that the universe is bathed in these fundamental sub-atomic particles. Other particles can interact with Higgs bosons. In so doing they are slowed below the speed of light and become heavy. The rate at which another particle interacts with the sea of Higgs bosons is roughly what determines the mass it acquires.

Higgs interactions are postulated to flip the handedness of the spin of the colliding particle. Spin is just one of several basic characteristics particles possess, others being mass and charge. Consider the tip of your thumb to mark the direction along which a particle is travelling. Now curl your fingers around – these show the direction in which the particle is spinning about the axis defined by your thumb. The handedness of the spin – whether it is clockwise or anticlockwise as you look down on the tip of your thumb – depends on which hand you chose.

Experiments have demonstrated that neutrinos always have the same handedness of spin. The implication is that they cannot interact with Higgs bosons, since this would require them to be flipped into a non-existent state. The neutrino masses had therefore been set to zero in the Standard Model. The results from the solar and atmospheric neutrinos, and helioseismology, discomfort this mass-free picture. Neutrinos appear to be able to change flavour – to change from something into something else. The interactions involved imply they must be slowed to beneath the speed of light. Neutrinos must therefore possess mass in order to be able to oscillate between states.

The Standard Model needs to be extended to resolve this conundrum. Massive neutrinos also have implications for cosmology. The universe is awash with neutrinos. The amount of missing mass they contribute depends upon the mass of the neutrinos, in particular that of the heavier muon flavour, and the total number of particles. The results from the Sudbury Neutrino Observatory have placed limits on the total mass all flavours might contribute to the universe. Current prejudice suggests they may only provide a tiny fraction of the total missing mass. This is probably insufficient to have an important influence on the evolution of the universe. That said, better data are still needed to place tighter constraints on the properties of the neutrinos, and their contribution to the missing mass. As noted by John Bahcall,[14] the Super-Kamiokande and Sudbury results have rested on observations of only a tiny fraction of the total number of neutrinos generated in the Sun. These neutrinos come from the high-energy part of the neutrino spectrum. There is therefore a clear need to extend the data collected on the different flavours into the more plentiful, lower-energy part of the solar neutrino spectrum. A whole new generation of neutrino observatories are being prepared for this task. Helioseismology also has an important part to play in pinning down further the interior structure of the Sun. But new surprises continue to be thrown up, as Chapter 11 will demonstrate. The moral? We cannot rest on our laurels.

---

[14] Overviews of the solar neutrino problem and its historical development are given in: J.N. Bahcall, *Physics Reports*, 333, 2000, p. 47; and J.N. Bahcall, *Nuclear Physics B Proceedings Supplements*, 118, 2003, p. 77.

9

IN A SPIN

Until the advent of helioseismology, measurements of the rotation of the Sun's interior did not exist. Theories for what the internal profile ought to look like had been developed. These theories were based on observations of the rates at which the surface of the Sun, and other stars, rotated, together with knowledge of the behaviour of spinning fluids, and prejudice regarding the then accepted understanding of the Sun's internal structure. The rotation of the interior was, and remains, a difficult problem – the rotation rate can vary in depth and latitude because the Sun is not a solid body.

Science is often at its most challenging and exciting when observation conflicts with theory, when the real picture that emerges throws up details that are wholly unexpected. This is what happened when helioseismology uncovered the spinning Sun. There were three big surprises. Two related to what was found within and just beneath the convection zone, the other to what was revealed deeper down. All have led to important revisions of our views on internal rotation in stars.

The story of these discoveries is spread over the next two chapters. We concentrate upon the deep-lying layers in this chapter – the radiative interior and core – and work our way out to examine the base of the convection zone and above in the next.

Let us begin with a general discussion of rotation in stars and the importance of angular momentum. Rotation is a common feature of astrophysical structures covering a vast gamut of scales in length and mass.
The introductory remark in a recent review by Mike Thompson, Jørgen
Christensen-Dalsgaard, Mark Miesch and Juri Toomre sums things up
quite nicely: 'Rotation is ubiquitous in astrophysical systems.'[1] Clusters,
galaxies, stars, planets and moons all show rotational behaviour, and
with it they possess angular momentum.

When we think of the momentum associated with a body we often
picture something moving in a straight line, like a car on a road. This is
linear momentum. Angular momentum is associated with motion
around a pivot point – for example that of the rotational motion of a
child on a swing; of a wheel about its axle; of a satellite that spins about
one of its axes to maintain stability; or of a star about its axis of rotation.

Keeping track of the angular momentum contained within a system
provides important clues to the behaviour and evolution of the
system. Sometimes, momentum can be exchanged between parts of the
system, altering radically the characteristics of the constituent elements.
An example is when a compact, extremely dense neutron star accretes
material off its nearby binary-star companion. Exchanges of this type
often take place in such a way that the total angular momentum of the
system remains little changed. Angular momentum is conserved. It is
important to know when, and when not, this holds true.

The impact of angular momentum conservation is often illustrated in
physics textbooks by the example of a spinning figure skater. Here, we
use a space age take on the same idea.

An astronaut is on a space walk testing a new thruster backpack. The
pack malfunctions, leaving a solitary thruster turned on. Eventually the
supply of fuel to the thruster runs dry but not before the astronaut has
been sent into a dizzying spin. The rotation takes place about the long

---

[1] This review gives an excellent, in-depth look at the internal rotation of the Sun:
M.J. Thompson, J. Christensen-Dalsgaard, M.S. Miesch and J. Toomre, *Annual Reviews
of Astronomy and Astrophysics*, 41, 2003, p. 599.

axis of the astronaut's body; a certain amount of angular momentum is associated with this motion. This then is the system whose properties we seek to investigate. Because we assume nothing else affects it, we say the system is 'isolated'.

In an attempt to slow the rate of rotation our astronaut extends their arms and legs. The distribution of mass within the system changes, a larger fraction now being a greater distance from the axis of rotation. This means the moment of inertia of the astronaut–backpack combination has increased – more energy would be needed to keep the astronaut spinning at the same rate. However, the angular momentum of the system remains unaltered. This is because no other object or body has exerted an external twisting force (torque) on the astronaut to alter their spin and there is no other mechanism, like friction, which might damp out the motion. Since the product of the moment of inertia and angular velocity gives the angular momentum of the astronaut, the rotation rate must decrease if the momentum is to be conserved.

Provided the backpack were not too heavy, this would probably have a noticeable effect that would allow the astronaut to regain their bearings. On the other hand, were one of their colleagues to come to their aid, exerting an external force to stop the rotation, the angular momentum associated with the motion of the astronaut would be transferred to the colleague. The system would no longer be isolated.

The idea that a large change in a body's moment of inertia can have consequences for its rotational behaviour plays an important role in stellar evolution. The way in which angular momentum is exchanged and altered during the various stages is complicated and in some cases poorly understood.

Stars form from the extremely cold clouds of hydrogen and helium found in the interstellar medium. The event that triggers the initial stage of star formation is when the denser parts of the cloud begin to collapse. For this to happen, the forces of mutual gravitational attraction of the component molecules must win out over those arising from the kinetic pressure of the gas. The former seeks to pull the molecules towards one another – to collapse the cloud – while the latter works against this.

The cloud eventually fragments into smaller, collapsing clouds. Provided the fragments can continue to cool quickly, so that any temperature variations are quickly smoothed out, conditions will be favourable for collapse to continue. The rate of collapse is reined in when this is no longer the case. Clouds become opaque in the infrared part of the spectrum and so their radiation finds it hard to escape. Excess heat is no longer so readily exchanged with the surroundings and the collapse is eventually halted.

As stars form from molecular clouds they inherit the angular momentum of the clouds. This may derive originally from turbulence in the interstellar medium or from differential rotation exhibited by the galaxies within which star-forming clouds are abundant. Either way, a modest amount of rotation is greatly amplified when a cloud collapses to form a star. This is because the moment of inertia decreases by many orders of magnitude as the cloud shrinks, whereas angular momentum is approximately[2] conserved during the process. The young stars that result rotate much more rapidly than their cloud progenitors.

This pattern of behaviour is a common one. Many of the structures observed in the universe formed from the collapse of larger, less dense structures. From only a hint of rotation in the progenitor, a much stronger pattern is likely in the daughter structure.

A similar, very large change in inertia takes place during the final stages of the evolution of stars, when they lose their outer envelopes but very little of their total mass. What remains is in some cases a neutron star, an extremely dense, compact object only a few kilometres in diameter. If angular momentum is approximately conserved this demands that the rotation rate of the neutron star be many orders of magnitude larger than that of the much larger star from which it evolved. Observations of pulsars bear this out.

---

[2] A small fraction of the angular momentum is transferred elsewhere.

These are neutron stars that emit directed beams of radiation from their magnetic poles. The magnetic axis is not aligned with the rotation axis. The effect of a favourable inclination with respect to the Earth is for the beam of radiation to be swept across the line of sight by the rotation like the beam of a lighthouse. The separation in time of the pulses of radiation observed then provides an estimate of the neutron star's rotation period. The periods range from just over a millisecond up to a few seconds – compared with the several- or many-day periods normally associated with Sun-like stars.

So, the early and latter stages of stellar evolution exhibit marked changes in rotational behaviour. But what happens in between?

The collapse of clouds during the early phases of stellar evolution is such that the inner parts of the cloud collapse most rapidly. As the cloud fragments take on their own identity, protostars begin to form with dense cores and tenuous outer envelopes. As the inner layers collapse, the resulting conversion of gravitational to thermal energy sees the temperature in the core of a protostar steadily rise.

Things are more sedate in the outer regions. Here, material continues to be accreted by the core, but temperatures are rather lower. As the outer layers become opaque to radiation another means must be found to carry the energy being released by the contraction to the surface. Protostars therefore become convective.

Since collapsing clouds rotate, the gravitational collapse is easy along the polar axis but is inhibited in planes at right angles to the rotation axis, because this is the direction along which centrifugal force acts. A disc of material may therefore be formed surrounding the star (a circumstellar disc).

Eventually, temperatures in the innermost parts reach a few million degrees centigrade. This heralds a coming of age for the protostar. It can now convert hydrogen into helium by thermonuclear fusion in its core. The copious amounts of heat thereby released provide a source of pressure that supports the protostar against the forces of gravitation. The collapse is halted and the newly formed, fully fledged star finds itself in a state of hydrostatic equilibrium, the various forces at play balancing one another. It has become a Main Sequence star.

This is a simplified picture of the changes a star goes through during its formation, but it serves to illustrate the salient points of interest to us here. The contraction from cloud to star will have resulted in a large increase in the rate of rotation. If a circumstellar disc is formed, angular momentum can be exchanged between it and the star, thereby influencing the latter's dynamic evolution.

Observations of young clusters of stars – whose members are all assumed to have formed at about the same time – suggest that Sun-like stars exhibit a wide range of surface rotation periods at this stage. However, if evolutionary time is wound forward, by making observations of older clusters, two things become clear. First, surface rotation rates in these more evolved objects are somewhat lower than in the younger clusters, meaning Sun-like stars spin down as they get older. Second, the spread in the observed rates is much less pronounced than at the earlier epoch, meaning some mechanism must act to make the surface rotation converge on similar, slow rates. How might this be achieved? The accepted paradigm goes something like this.

Sun-like stars have a deep radiative core and interior and an outer convective envelope. The segregation of the star into this structure is followed shortly after by the onset of dynamo action in the outer layers. This dynamo amplifies the primordial magnetic field in the star – assumed to have a simple dipole form, like a bar magnet – and re-shapes it into a series of much more elaborate magnetic structures. We shall look at the solar dynamo in greater depth in the next chapter. Here, it is important to recognise the dynamo does its work in the sub-surface convection zone, current opinion placing the seat of the dynamo just beneath the base of the zone. Magnetic structures arising from the dynamo permeate the outer layers and atmosphere of the star.

When the conductivity in the solar plasma is very high, the plasma and lines of magnetic field – the same lines traced out by iron filings scattered on paper around a bar magnet – are 'frozen' together: the gas finds it hard to cross lines of magnetic field and the gas and field act like they are tied to one another. The relative sizes of the magnetic and kinetic gas energies then determine which drags the other around. This 'frozen field' effect gives rise to a magnetic brake that slows the rotation of the outer parts of the star. To understand how this works we must again think in terms of angular momentum.

Consider a spherical spinning and expanding shell of material. If no external forces affect the rotation – so we may assume angular momentum is conserved – the expansion will be accompanied by a slowing of the rotation. This is because the moment of inertia of the shell increases with the passage of time as the shell gets bigger. Now consider the case where some external agent enforces rigid rotation. We can no longer invoke the conservation of angular momentum. The action of the external force will now cause an increase in the angular momentum of the shell as the latter expands.

This second case is relevant to the outer atmosphere of a Sun-like star. The Sun is continually losing material from the corona via the solar wind. So there is an outflow, or expansion, of material in its atmosphere. We may approximate the lines of magnetic field in the corona as being radial so that they point in a direction at right angles to the surface of the star. Remember that the stellar matter and magnetic field are tied together. As the star rotates, the lines of field therefore act to shepherd a similar rigid rotation of the expanding plasma. The field in effect provides the extra torque alluded to in the example discussed above. This means that the angular momentum per unit mass of the material increases as the material moves out through the atmosphere. The source it must tap into to acquire this extra angular momentum is that stored in the rotating star.

The separation between the radial lines of field increases further from the surface of the star. Since the concentration of these lines determines the strength of the magnetic field, the field strength must decrease at greater radii. Indeed, it will do so to such an extent that at some critical radius the field can no longer control the plasma. Beyond this point the stellar material can be lost, carrying away with it some of the star's angular momentum. Over time the action of this magnetic brake, via the continual loss of angular momentum, slows the rotation of the outer layers.

Rotation is a key part of the dynamo. Take two stars whose convection zone properties are similar, but one rotates more rapidly than the other. Observations show that faster rotators tend to have a more vigorous dynamo. A strong dynamo implies a strong brake. So, the faster star in our pair will develop a more effective magnetic brake than its more slowly rotating counterpart. Over time the surface rates of rotation of the two stars will converge as they both slow down.

One can now see why the magnetic brake idea is so attractive. It provides a neat explanation for the surface rotation observed in clusters of widely differing evolutionary ages. But this still says nothing about the more deep-lying radiative layers beneath the convection zone.

Stellar dynamos are assumed to operate in the outer layers, such that only these parts are affected directly by the action of the magnetic brake. Conventional wisdom regarding the core and deep radiative interior suggested they would be expected to rotate much more rapidly than the layers above. The final stages of pre-Main-Sequence evolution are when a Sun-like star would have developed its radiative interior. Whereas the outer layers would have come under the influence of the magnetic brake, the newly forming core would have contracted a little more, leading it to rotate even more rapidly. The result? A slowing rate of rotation in the envelope, and a rapidly spinning core.

The question then arises of the degree of coupling between the radiative interior and outer layers. Coupling allows the envelope to draw on the large reservoir of angular momentum residing in the core. This has two consequences. First, it will delay the rate at which the envelope is spun down. And, second, it will bleed momentum from the core, thereby altering the rotation in the deep interior. The extent to which the core and envelope are coupled therefore plays an important role in a star's dynamic evolution.

In the pre-helioseismic era models of and conjectures about the deep interior lacked any *direct* observational evidence to prove or disprove them. The field therefore awaited the first estimates of the rotational frequency splittings of the core-penetrating modes with some anticipation. Besides answering questions posed by stellar evolution theory, the data would allow helioseismology to constrain one of the well-known tests of Einstein's general theory of relativity – the advance in the perihelion of Mercury.

This test is based around matching observations of subtle changes that occur over time in the orbit of the planet Mercury with predictions from theory. These tiny variations have contributions from several sources. The two of interest here come from the exotic effects of relativity theory and the Sun's departures from a pure spherical figure. Helioseismology was in a position to say something precise about the Sun's shape. It could therefore potentially pin down the small solar contribution to allow scientists to determine the remainder attributable to relativistic effects. The question was then which relativity theory – Einstein's or a competitor's – provided the best match to what was left over?

The history of this test is interesting. Let us start by saying something about the 'subtle changes' in Mercury's orbit. This planet follows an elliptical path around the Sun. The long, or major, axis of the ellipse joins two points: the aphelion, or distance of furthest approach from the Sun, and the perihelion, or distance of closest approach. These distances are approximately in the ratio 1½:1.

The direction of the major axis changes over time with respect to the stars. It precesses through a small angle after each successive orbit. One way to measure this is to track the change in the direction of the perihelion. The test in question is therefore based on the resulting advance of Mercury's perihelion. The advance covers an angular extent of approximately 1.5 degrees on the sky every thousand years. Although the effect is small, it is measurable.

Modern estimates are made from radar-ranging data. A careful series of observations with a telescope can also do the job (although less precisely). This is how the French astronomer, Urbain Jean-Joseph

Leverrier, first obtained a description of the orbit of Mercury and its perihelion advance. He made an initial stab in 1843. By 1859 he had much better observations and an intriguing new solution.

Several factors contribute to the precession. The largest comes from the 'classical' influence of the other planets in the solar system – gravitational effects that can be calculated using straightforward Newtonian mechanics. Venus accounts for just over half the combined effect; the next largest contribution comes from Jupiter. When Leverrier calculated the size of this classical contribution he found it fell short of the observed value of the perihelion advance. Modern observations indicate the deficit is about eight per cent of the size of the classical prediction. This mismatch is termed the 'anomalous advance of the perihelion of Mercury'. How did it arise?

Two options were given initial consideration. First, astronomers of the time had to entertain the possibility that their estimates of the other planets' masses might be in error. The masses were needed to determine accurately the size of the classical contribution. A second, more exciting (and less uncomfortable) proposition was the impact of an as yet undiscovered planet – given the name Vulcan by another French astronomer, Babinet – or a ring of planetoids or asteroids. Observations failed to uncover evidence of either. It was not until much later in the nineteenth century that alternative solutions were considered.

Two offered by the American astronomer Simon Newcomb turned out to have greater relevance. One concerned the shape of the Sun. He pointed out that an additional Newtonian contribution to the advance could arise from the impact of a distorted solar gravitational potential.

All previous calculations had assumed the Sun was a perfect sphere. This would indeed be the case if rotation were absent and important characteristics then had spherical symmetry. We know the Sun is supported against its own weight principally by the gradient of gas pressure in its interior. In our spherical scenario, contours of constant pressure would lie on spherical surfaces and also coincide with contour lines mapping the gravitational field of the Sun – the so-called gravitational potential.

Rotation can flatten the Sun, because the Sun is made of compressible gas. The more rapidly the Sun rotates, the flatter it will become: the distance between its poles will be shorter than the distance between the extreme edges at the equator. The Sun then has an oblate figure, the oblateness being a measure of the fractional difference in the equatorial and polar diameters. Surfaces of constant pressure, and crucially those of constant gravitational potential, are no longer spherical in an oblate Sun.

Newcomb recognised a distortion of the gravitational potential would alter subtly the forces the Sun exerts on Mercury. Newcomb was therefore the first scientist to raise the possibility of an oblate Sun *and* a rapidly rotating interior to account for the oblateness.

For his second suggestion Newcomb made the more drastic step of considering tweaks to Newton's Laws of gravitation. The ideas in what he proposed turned out to be a dead end. However, the philosophy of challenging the basic Newtonian predictions foreshadowed the more profound explanation that would soon gain general recognition. This required a whole new way of looking at how bodies interacted with one another.

Einstein's general theory of relativity sought to explain the effect of gravitational fields on space and time. An important prediction from the theory was a small contribution to the perihelion advance of Mercury, arising from the effects of the curvature of space-time (different from the Newtonian case). Provided the oblate-Sun contribution was negligible, it turned out that the predicted relativity advance accounted very nicely for Leverrier's 'anomalous' advance, bringing observation and theory into line (Einstein's first test of his new theory). Throughout the first half of the twentieth century this was the accepted result – the anomalous advance was a relativistic effect, and Einstein's general theory of relativity could account for its size.

However, the years that followed brought reinvigorated interest in the problem. This owed in part to the development of competing relativistic theories, conspicuous among these being the theory of Carl Brans and Bob Dicke.[3]

[3] C. Brans and R.H. Dicke, *Physical Review*, 124, 1961, p. 925.

Different theories could predict different-sized relativistic contributions to the advance. But how was one to decide on the correct value? To admit a change in the size of the relativistic contribution meant something else had to give; the total prediction still had to match the observations. The obvious candidate for change was the distorted-Sun contribution. If it could be shown the gravitational potential of the Sun was non-spherical in shape, there might be a non-negligible contribution to the perihelion advance from the potential. This would drive down the size of what was left over for the relativistic part. With enough oblateness, a conflict with Einstein's general theory of relativity might be the result.

Brans–Dicke theory could be tuned: instead of producing one hard-and-fast estimate for the relativistic contribution, it could predict a range of values dependent on the choice of a particular parameter in the theory.[4] When this constant was very large the predictions from Brans–Dicke looked like those from Einstein. When it was smaller, the two began to diverge, since the Brans–Dicke prediction got smaller also. Brans–Dicke could therefore come to the rescue *if* the Sun turned out to be very oblate.

Securing an accurate and precise determination of the shape of the Sun's potential would be no easy matter. It would be necessary to measure the rotation profile throughout the Sun's interior to establish whether the rotation really did cause a significant distortion. Our historical sketch has now only reached the 1960s, and there was then no way to probe the internal rotation. That all changed with helioseismology.

Helioseismology proved able to measure the rotation profile and resolve the issue. However, the data needed to accomplish the task only became available in the early 1980s. In the meantime the method that was

---

[4] This fixed the effectiveness of the long-range field that the theory postulated was generated by matter. This field, calibrated by the gravitational constant *G*, was proposed as a means of providing the origins of the inertia of matter.

pursued involved attempting to determine the distortion of the gravitational field by measuring the Sun's oblateness.

It can be shown that several factors contribute to the oblateness. There are contributions from velocity and magnetic fields present in the surface layers of the Sun. And there is the sought-for contribution from the distorted gravitational potential, which would carry the signature of any rapid internal rotation.

The distortion of the potential is commonly expressed in terms of a gravitational quadrupole moment (which would be zero in the spherically symmetric case). A simple multiplicative factor relates the quadrupole moment to the surface oblateness given by the distorted potential: measure this part of the total oblateness and you can estimate the contribution to Mercury's perihelion advance.

This is, however, no easy task. The measurement presents many technical challenges – it is not like locating the edge of a solid object. Even when a (hopefully) robust estimate of the total oblateness has been obtained, the measurement must be properly corrected for the unwanted surface contributions. The largest of the unwanted effects comes from the rotation in the near-surface layers. Observations of the surface rotation can be used to estimate the size of the effect; when the resulting value is subtracted from the total oblateness, what is left over provides an estimate of the sought-for gravitational quadrupole moment. The correspondence is only exact when the regions being observed are free of magnetic fields or velocity flows that depart from the large-scale pattern of rotation. So, more care needed.

Bob Dicke was one of the scientists to deal with these issues at some length. The alternative theory of relativity he had formulated with Carl Brans gave him an obvious interest in the problem. Dicke was not only an excellent theoretician; he was also a talented instrumental physicist. He determined to measure the oblateness of the Sun and, together with Mark Goldenberg and in the early stages Henry Hill (then also at Princeton), he began the construction, in 1963, of an instrument designed specifically for the purpose. After a series of careful tests, and updates in design, the instrument gave its first result, published in 1967 by Dicke and Goldenberg – an oblateness of one part in twenty thousand.

This value appeared to have severe consequences for Einstein's general theory of relativity. Its size suggested the distorted-Sun contribution was not so negligible after all. It would amount to just under ten per cent of the anomalous advance of Mercury's perihelion. This implied that only about ninety per cent of the anomalous part could now be set aside as relativistic. Predictions from Einstein's theory needed pretty much the full hundred per cent.

Dicke and Goldenberg's result also had important implications for the Sun's interior structure, for it suggested a rapidly spinning core might be needed to account for the surface oblateness. The concept of rapid internal rotation also happened to be in vogue at the time for another reason – it offered a possible way to solve the solar neutrino problem.

In the presence of rapid internal rotation thermal pressure would no longer be required to carry the full burden of support to maintain the star in equilibrium: temperatures in the core could be lower than previously thought, provided the contribution from centrifugal force was non-negligible. Roger Ulrich (in 1969) and Ian Roxburgh (in 1974) were two scientists who considered the ramifications of rapid internal rotation. Also, in 1973, Pierre Demarque and colleagues presented a series of solar models with rapidly rotating interiors, which predicted surface oblateness values ranging between about one part in forty thousand and one part in six thousand. Dicke and colleagues' measured value lay safely within this range.

Dicke and Goldenberg's result was, however, controversial, and matters were made no clearer when Henry Hill (now based in Arizona) and Tuck Stebbins offered an independent measurement of the oblateness which was much smaller.[5] This was the state of play when

---

[5] It was from his observations of the solar diameter that Hill claimed the discovery of global oscillations (see Chapter 6).

helioseismology was about to offer its first estimates of the internal rotation.

The rotation of the Sun splits the modes into several components (Chapter 3). Getting to grips with the internal rotation, through the use of the p modes, is therefore a matter of measuring these splittings – the separation in frequency of the various components of each mode. One has to be aware that rotation is not the only thing that can contribute to the sizes of the splittings. For example, magnetic fields and flows of material can add their effects too. However, the rotation gives a nice, even, symmetric pattern. If you can separate out the symmetric part of the splitting from the rest, you can infer the rotational contribution.

To obtain an estimate of the rotation in the deep interior requires splittings from the low angular-degree modes that penetrate these layers. However, to make effective use of the low-degree splittings, splittings from intermediate and high degrees are also needed.

The splittings of the low-degree modes are made up of a suitably weighted contribution of the rotation throughout their cavities – from the core all the way up to the surface. The rotation in the outer parts of the interior therefore needs to be properly pinned down, courtesy of the shallower-penetrating, higher-degree modes, to allow the rotation deeper down to be inferred by the inversions. (The same is true for the sound-speed inversions.)

The task of uncovering the rotation in the deep interior is much tougher than uncovering the sound speed. As we shall see below, the splitting of a typical low-degree mode is a few thousand times smaller than its frequency. The result? It is not possible to measure the splittings to anything like the same fractional precision achievable for the frequencies – and, as one might expect, the effects of this carry over into the rotation inversions. Although inferior, the precision in the splittings is still good enough to allow scientists to peer into the spinning radiative interior.

We know the modes spend little time sampling the core – the low sensitivity means the rotation in the core contributes only a small fraction to the total splitting of a mode. How these small contributions compare in size with the uncertainties on the measured splittings is therefore crucial in determining the effectiveness of the probe of rotation offered by the sound waves. Take the first types of modes to be split, the dipole ($l = 1$) modes.

When we consider in detail how one of these modes – say one near the centre of the oscillation spectrum – senses the rotation, we find that only about eight per cent of its total splitting comes from the rotation in the core regions (here defined to be the volume contained in the innermost twenty-five per cent by radius). Rotation in the radiative zone, between twenty-five and seventy per cent by radius, contributes roughly twenty-nine per cent to the splitting. This takes us to the base of the convection zone, so that the largest contribution to the splitting – the sixty-three per cent or so left over – comes from the outer layers above.

These numbers suggest that if one wishes to uncover the rotation of the deep radiative interior and core, one will face difficulties if the fractional precisions in the splittings are greater than a few per cent.

Before the first helioseismic measurements, scientists were able to estimate the magnitudes of the splittings they might expect to see on the basis of observations of rotation at the surface, and predictions made from models of rotating stars.

The surface observations showed that the surface rotated, on average, about once every 27 days. Let us suppose the whole interior of the Sun rotated at this period. The splittings of the modes would all be roughly the same,[6] their size reflecting this period. To turn the 27-day period into a frequency is a simple matter converting the value into seconds and then taking the inverse to give a frequency in Hertz. The value one gets is

---

[6] Because of effects of Coriolis and centrifugal forces the splittings would differ slightly.

0.4 micro-Hertz.[7] This is the splitting that would separate adjacent peaks of modes in our simple scenario.[8]

To measure such a tiny splitting, one needs to be able to master two things – the frequency resolution of the data, and the widths of the mode peaks. The first one can control. By collecting more data, the resolution in frequency will improve in proportion to the length in time of the dataset.[9] Width is altogether another matter – of particular concern for the Sun-as-a-star observations.

The intrinsic widths of the mode peaks provide a measure of the damping and are fixed by what happens on the Sun. They are something one simply has to live with. When widths become comparable to the splittings, it is not hard to see that one is going to run into difficulties. This is indeed a problem for the higher-frequency modes in Sun-as-a-star data (see Figure 9.1). Their resonant peaks are so wide that they overlap considerably.

The problem can be neatly side-stepped for observations that resolve the surface of the Sun into many pixels. With these more complicated data it is possible to separate the different split components into different oscillation spectra.[10] This at once reduces the congestion of nearby peaks.

---

[7] A small correction also has to be added to the raw splitting to get the actual rotation rate in order to allow for the observers' motion with respect to the Sun. For observations made from a vantage point that orbits the Sun with a period of 1 year the frequency correction has a magnitude of the inverse of 1 year – about 0.03 micro-Hertz. Because the Earth's orbit about the Sun is slightly eccentric the correction for ground-based observations also varies with the time of year (this is so small it can be disregarded).

[8] That is, adjacent in the spherical harmonic azimuthal degree, m. Some peaks are missing from the Sun-as-a-star data. For observations made in or close to the ecliptic – e.g. Earth-bound or from L1 – modes having odd $l + m$ leave a negligible signature in the observations.

[9] When there are gaps in the data – for example those given by a ground-based network – the effective resolution is slightly poorer.

[10] Each spectrum then contains the more widely separated overtone peaks of a given l and m. In reality other modes leak into the spectrum; only with a completely global view of the Sun would this problem go fully away (for example using several suitably positioned spacecraft).

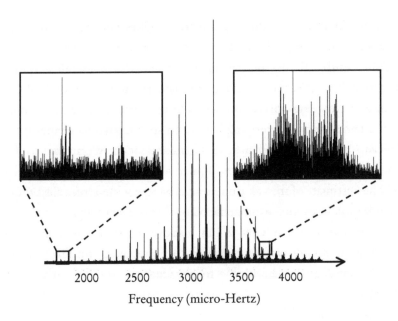

Figure 9.1. The low-degree Sun-as-a-star spectrum (here made from several years of BiSON data); two narrow slices of it are highlighted in the panels. Both slices have been selected to show pairs of $l = 2$ (towards the left-hand side of each panel) and $l = 0$ (towards the right-hand side) modes. The $l = 2$ mode is cleanly split by rotation into three observable components in the low-frequency inset (two are missing owing to the geometry of the observations). The peaks are far wider in the right-hand panel, taken from a higher-frequency part of the spectrum: the result is that individual peaks of the $l = 2$ mode are blended together. Courtesy of the BiSON team.

In our simple numerical example above we assumed an interior whose rotational properties were the same everywhere – a rigidly rotating Sun. A reasonable question to then ask is: by how much would the 0.4-micro-Hertz splitting change if we were to speed up parts of the interior? Let us paint a simple picture in which we keep most of the interior rotating at 27 days but allow that part contained in the innermost twenty-five per cent by radius to spin up. This kind of step-wise change in rotation is admittedly not what one would expect to see inside the Sun – but it suffices to show the size of effect involved and is a straightforward calculation to perform.

First, let us assume the rotation rate doubles in the core, so it has a period of 13½ days there. This does not mean the overall splitting doubles in size. Remember, the core contributes only a small amount to the total. If we take the dipole mode considered above, the total splitting would increase by only 0.03 micro-Hertz. With the core rotating five times faster than the overlying layers, the splitting would change by 0.1 micro-Hertz from 0.4 to just over 0.5 micro-Hertz. With a twenty-fold increase, it would rise to around 1 micro-Hertz. This last splitting gives a rough estimate of the value one would have expected from the types of rapidly spinning core fashionable before helioseismology.

The changes are not huge, but because the splittings of the core-penetrating modes can be measured to a precision of a few per cent, helioseismology has been able to elucidate the rotation of the deep interior.

The first results of using p mode splittings from the 5-minute spectrum arose out of a collaboration, in the late 1970s, between Ed Rhodes, Roger Ulrich and Franz Deubner. As early as Christmas 1976, Ed had been pondering the use of his data to look at the rotation in the near-surface layers. Franz had also been thinking about these issues and had realised that data could be used to probe the rotation in different layers near the surface. When the three scientists teamed up they analysed splittings of high-wavenumber modes; the papers that resulted from this work provided the first tantalising glimpse of the rotation in the near-surface layers of the interior.[11]

The first attempts to say something about rotation much deeper down pre-dated this. In the 1976 paper in which they reported on the 160-minute 'oscillation', George Isaak and his colleagues in Birmingham made a first stab at using the apparent absence of splittings in the peaks to infer an upper limit on the rotation throughout the interior. Given the controversial nature of these data, the upper limit they found – of about

[11] The first to appear was E.J. Rhodes, Jr, R.K. Ulrich and F.-L. Deubner, *Astrophysical Journal*, 227, 1979, p. 629.

12 hours – must be viewed with some caution. But George and colleagues had appreciated what split modes could offer, so that when they uncovered the whole-Sun nature of the 5-minute modes measurement of the splittings became one of their first priorities.

The average splitting they initially uncovered – a result published[12] in 1981 – was about 0.75 micro-Hertz. This number was higher than the observed surface rate and therefore seemed to point to a faster spinning interior. George did some calculations and concluded the deep interior of the Sun was rotating anywhere from three to nine times faster than the observable surface. The precise value depended upon the size of the rapidly spinning central region. George noted that the uncovered splitting fell 'far short' of the rate required to give a large oblate figure at the surface. The result appeared to contradict Dicke's oblateness measurements and provided some reassurance for supporters of Einstein's theory.

A similar conclusion was reached a year later by Douglas Gough and Henry Hill. In new data, Hill claimed to have found not only continued evidence for p modes at low frequencies, comfortably below the 5-minute frequency region, but also some g and f modes and – crucially – splittings. He was able to convince Douglas, and the two began to work together to identify the modes, and the sizes of the splittings. Subsequent disagreement over how the data should be interpreted led them to go their separate ways on the project, and Douglas and Henry each published their own paper.[13] Although much of the detail contained in these differed, the main conclusions did not. The splittings, if real, were consistent with Sun's distortion making a negligible contribution to Mercury's perihelion advance.

The consensus view now is that Hill's data most likely did not contain the claimed modes (and splittings). The field was anyway able to place

---

[12] A. Claverie, G.R. Isaak, C.P. McLeod, H.B. van der Raay, T. Roca Cortés, *Nature*, 293, 1981, p. 443.
[13] D.O. Gough, *Nature*, 298, 1982, p. 334; H.A. Hill, R.J. Bos and P.R. Goode, *Physical Review Letters*, 49, 1982, p. 1794.

more store in the 5-minute observations, whose bona fides were confirmed by the fact several independent observations were in excellent agreement. The 5-minute peaks stood proudly above the noise in the various data and identification of the peaks was on very safe ground. Soon, 5-minute splittings needed to perform the first inversion arrived, courtesy of Jack Harvey, Tom Duvall and Ed Rhodes.

These data were collected at the Kitt Peak Observatory by the instrument with the ingenious cylindrical lens at the front. Measurement of splittings of the outermost mode components selected out by the lens permitted an estimate of the rotation in the regions of the interior close to the solar equator. Douglas Gough performed an inversion using the techniques he had developed in the early 1980s. The publication of the results[14] – with Tom, Jack, Phil Goode, Wojtek Dziembowski and John Leibacher – pre-dated by a year the first sound speed inversion (also made with Jack and Tom's data).

The conclusion was that the rotation in the equatorial layers persisted at the surface value right down into the radiative interior. The results for layers in the inner forty per cent, by radius, were less certain. There was a suggestion of a slight upturn in the rotation rate; the apparent value was substantially slower than predictions from the best rotation models of the time.

The uncovered profile also enabled a prediction of the distortion of the Sun's gravitational potential and its impact on the perihelion advance of Mercury. Although it was not possible to estimate the rotation in the deepest layers – there were simply too few of the lowest-degree modes in the data – a fairly robust estimate of the distortion was forthcoming.

The mathematical expression used to describe the gravitational quadrupole moment tells us the rotation in the core contributes rather less to the size of the moment than does the rotation in the radiative and

[14] T.L. Duvall, Jr, W.A. Dziembowski, P.R. Goode, D.O. Gough, J.W. Harvey, J.W. Leibacher, *Nature*, 310, 1984, p. 22; Douglas Gough had already published a discussion of the inversion technique and applied it to artificial data having similar characteristics to those observed (D.O. Gough, *Philosophical Transactions of the Royal Society of London*, 313, 1984, p. 27).

convective zones further out. So, provided there was not some horren-dously steep upturn of the rotation rate in the core – a range of plausible possibilities were allowed by the error estimates of Douglas and col-leagues – the data implied a tiny quadrupole moment and a tiny oblateness, of about one part in 200,000, suggesting that any contribu-tion to the Mercury test was negligible. The tuning parameter in Brans–Dicke theory was then inferred to be so large as to make every-thing consistent with Einstein's prediction. This conclusion has been borne out by the longer, higher-quality datasets collected since. Modern estimates of the oblateness fix its value at about one part in three million, implying a solar contribution to the anomalous advance of the perihe-lion of less than one-tenth of one per cent.

As new data that looked at splittings of modes other than the outermost components became available, it became possible to infer the depend-ence of rotation upon latitude in and just below the convection zone. In the meantime improvements in the application of the inversion tech-niques – by the likes of Douglas, Jørgen Christensen-Dalsgaard, Mike Thompson, Frank Pijpers, Takashi Sekii and Sarbani Basu – enabled helioseismology not only to take advantage of the improved data but to also make the inferences more reliable.

But what of the rotation in the deepest parts of the radiative interior and core? To answer this required better, longer sets of low-degree data and became a key objective of the Birmingham team (and BiSON), the Tenerife group and Eric Fossat's team in Nice (through IRIS). These and other groups worked consistently throughout the 1980s and early 1990s to get better estimates of the splittings.

When results from these low-degree analyses are plotted now, they reveal a fascinating trend. After the first estimate of the early 1980s – a value of 0.75 micro-Hertz – there is a steady reduction in the sizes of the uncovered splittings. By the beginning of the 1990s typical values ranged between about 0.5 and 0.6 micro-Hertz. Though ruling out a very rapidly spinning core the values still pointed to an upturn of some sort in the very deep interior. This changed, and the long-term trend bottomed out, in 1994.

BiSON had then become the fully fledged six-station network it is today. With the new automated observatories and better coverage worldwide, the quality of the oscillation spectra that could be produced took a big leap forward. The key ingredient was the ability to detect a greater number of modes at low frequencies.

We have extolled the virtues of these modes on more than one occasion in the book – and shall do so again here. The narrow peaks they give in frequency are ideal material for measuring splittings. The splittings can then be measured more precisely – the peaks are so narrow that the uncertainty in the value of the splitting must be very small. What is more, the pattern then appears much more clearly. The peaks do not overlap, as they do in the higher-frequency parts of the spectrum, and problems created by such merging are circumvented. At low enough frequencies the splitting can even be estimated by putting a ruler to a plot. (The actual analysis is a little more involved.) One can have more confidence in the splitting values themselves, because they are more accurate.

When Rachel Howe, George Isaak, Yvonne Elsworth and Roger New arrived at the big helioseismology conference of 1994 – hosted in California by Ed Rhodes – they brought with them the results of Rachel's analysis of the latest BiSON data. They were not originally listed to present a talk on this topic to the assembled meeting. However, when word of the result spread, it was decided a slot in the programme should be found.

The typical value of splitting Rachel had uncovered – now close to 0.4 micro-Hertz – implied the deep interior was actually rotating at a rate no faster than the surface. This was backed up by the more complete analysis she made on her return to Birmingham, which even suggested the rate in parts of the interior might be slightly slower than at the surface.

The resulting Elsworth et al. paper – submitted and published with Douglas Gough as a co-author in 1995[15] – marked, I think, a sea change

[15] Y. Elsworth, R. Howe, G.R. Isaak, C.P. McLeod, B.A. Miller, R. New, S.J. Wheeler and D.O. Gough, 1995, *Nature*, 376, p. 669.

in this area. Estimates made by other groups soon began to converge on the 'slow' value. This happened as a result of other high-quality data becoming available, from the GOLF, MDI and VIRGO/LOI instruments on the SOHO spacecraft, and new analyses of IRIS – all the sets could now show the cleanly split modes at low frequencies.

So, we come back to the mysterious downturn of the splittings from 1981 to 1994. Alas, the deep interior was not spinning down over time. The trend flattened out because better data, with superior frequency resolution, became available. The earlier datasets were shorter and generally of poorer quality. The sharp low-frequency modes were then so drowned out by noise as to be unusable, which forced analyses to make use of the wider, higher-frequency peaks. The resolution of the data was also inferior, which made it harder to disentangle those peaks that could be seen. The splittings the old analyses gave overestimated what was really there.

Since the later 1990s the excellent quality of the long modern datasets has put us in a position to appreciate more of the subtleties present in the data. One important spin-off has been a detailed understanding of some of the pitfalls outlined above. We now have a clearer idea of when, and when not, to trust the splittings.

With the very best data now available it is possible to observe with some confidence low-degree modes at frequencies close to 1 milli-Hertz. These peaks are more than one hundred times narrower than their more prominent compatriots near the centre of the p mode spectrum. Their splittings can be determined to a precision of close to one per cent. Even then we are having difficulty penetrating the rotational secrets of the core.

You may have noticed I have been careful in my use of language – the text is peppered with phrases like 'the deeper-lying layers' or 'the deep radiative interior' when 'core' might have done. This has been deliberate. We can now be fairly confident of the picture of rotation we have down to about twenty-five per cent by radius. But at smaller radii the uncertainties are very large indeed.

In spite of these difficulties, we can say that large parts of the radiative interior are rotating at a rate comparable to that in the near-surface layers, a result most unexpected. Models of rotational evolution in Sun-like stars have to take account of this. Results from helioseismology that speak to what is happening at the interface between the convection zone and radiative interior also have a major bearing on the problem, as we shall now see in the next chapter.

10

GETTING ACTIVE - THE
SOLAR CYCLE

The window into the Sun we have seen opened in preceding chapters has revealed a few surprises and enabled us to refine the pre-helioseismic portrait of the Sun's interior structure.

Changes to the equilibrium structures of the deep interiors of stars like the Sun take place on vast timescales. The burning of hydrogen fuel into helium ash takes several billion years. These long-term evolutionary effects leave their imprint on the solar oscillations by virtue of a snail's pace adjustment of the interior structure as the star sedately ages on the Main Sequence. The frequencies of the modes decrease slowly but surely. But they do so at a rate that makes any direct observation of the effect a task beyond current observations.

The frequencies of the lowest-degree modes of the Sun are undergoing evolutionary changes of approximately one-millionth of a Hertz every six million years. On a timescale more practical from a human observer's point of view – say ten years – the evolutionary change is reduced to about a millionth of one micro-Hertz. To attempt to measure such a change demands that mode frequencies be observed to a level of precision superior to this.

The most precisely determined frequencies are those of the long-lived, sharply peaked modes at low frequency. The best estimates we have currently at low degree have associated uncertainties of a few thousand millionths of a micro-Hertz. Clearly, this is nowhere near the level of precision required. But any helioseismologist can tell you that we *do* see systematic changes in the frequencies. These take place on an 11-year timescale and are about 400,000 times larger than the evolutionary prediction above. The variations are regular, not random. So what causes them?

Martin Woodard and Bob Noyes uncovered the first evidence of such changes in the mid 1980s.[1] They looked carefully at data collected by the ACRIM instrument on board NASA's Solar Maximum Mission satellite. They found that the frequencies of the lowest angular-degree, whole-Sun modes were higher in 1980 than they were four years later. The first year coincided with a high level of activity on the surface of the Sun – for example, elevated numbers of sunspots – whereas at the latter time activity levels were much lower. The modes appeared to be responding to the Sun's 11-year cycle of magnetic activity.

The shifts were, on average, about 0.4 micro-Hertz in size. This meant the frequencies of the most prominent modes were increasing by roughly one part in ten thousand between the activity minimum and maximum of the cycle.

The accepted paradigm, then and now, is that the solar cycle is a phenomenon associated with the outer layers, not the deep radiative interior, of the Sun. The regeneration and reorganisation of magnetic fields hold the key to what drives the cycle. The appearance of new magnetic field and the decay of old field, features associated with this, and any resulting changes to the underlying solar structure are therefore assumed to be limited to the outer layers and the overlying atmosphere in which striking manifestations of the activity, like flares, are observed.

The fractional mass of the Sun contained in its outer envelope is very small. This means readjustments to the structure of this part have little impact on the basic evolution of the star. The speed of sound near the

---

[1] M.F. Woodard and R.W. Noyes, *Nature*, 318, 1985, p. 449.

surface is also much lower than in the deeper interior. The acoustic waves forming the modes therefore spend a greater fraction of their time near the surface, so they are more sensitive to changes there than to changes deeper down. A modest physical change at the surface, of little signifi-cance from a gross evolutionary point of view, therefore suffices to alter the frequencies at readily observable levels.

It is by detecting, and measuring in detail, solar-cycle changes – which now include variations not only in frequency but also in the likes of mode splitting, power and damping – that helioseismology is able to shed light on the processes responsible for the cycle and the physical changes that occur in the near-surface regions.

In this chapter we shall look at how a dynamo is believed to generate the solar activity, and discuss the insight that helioseismology has given into its workings. The patterns of rotation uncovered in and just below the convection zone have been particularly instructive in this regard.

Besides their illumination of our understanding of these processes there are other reasons for pursuing such knowledge. Eruptions of high-energy particles, associated with active phenomena like the solar wind and coronal mass ejections, can reach the Earth and leave their mark. When charged particles from the Sun arrive they interact with the Earth's magnetic field to create huge electric currents and geomagnetic storms. In March 1989, particles released by the Sun set off a storm that created havoc for navigation systems – compass bearings where thrown off by several degrees. The Quebec power grid in eastern Canada tripped out, leaving several million people without electricity for several hours. Other examples of interference include disruptions to radio communi-cations and occasional damage to satellites. An ability to predict reliably the occurrence of events like this would clearly be useful.

Another area of interest concerns longer-term variations in activity. During the latter half of the seventeenth Century sunspots – whose presence and variation in number are conspicuous examples of the surface activity and its cycle – all but disappeared from the solar surface

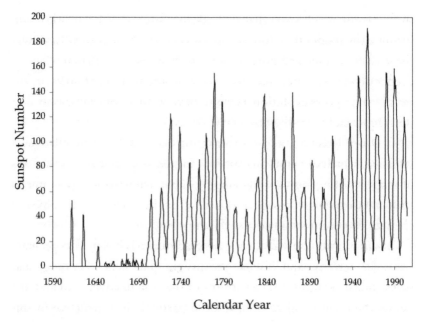

Figure 10.1. Variation of the sunspot number over the last 400 years. The period in the second half of the seventeenth century is called the Maunder Minimum. Data from the National Geophysical Data Centre.

(Figure 10.1). This period is called the Maunder Minimum.[2] Not only were these outward manifestations of the activity greatly reduced, but it is believed the total solar radiation output (the luminosity) may have dropped also.

The radiation output is measured by the so-called 'solar constant'. This is not quite constant but varies with the 11-year cycle, being larger at the peak of the activity, albeit by only a tenth or so of one per cent. The solar constant also varies on the timescale of the solar rotation, at a level

---

[2] It carries the name of E. Walter Maunder, who at the end of the nineteenth century (as superintendent of the Solar Department at the Greenwich Observatory) undertook a detailed study of the historic records relating to the phenomenon. This included a summary of the work of Gustav Spörer, who had first pointed out the apparent absence of sunspots. It was not until John Eddy revisited the observational evidence in the 1970s that the reality of the Maunder Minimum received widespread acceptance (J.A. Eddy, *Science*, 192, 1976, p. 1189).

of about two-tenths of one per cent. An extensive record of observations of the solar constant is available from several satellites, most recently from Claus Fröhlich's VIRGO package on the SOHO spacecraft.

A variety of studies have speculated that the total solar radiation may have been lower during the Maunder Minimum, by anywhere from one-tenth to three-tenths of one per cent compared with modern levels at the low point of the activity cycle.[3] There are implications here for the solar input to terrestrial climate. (It should be pointed out that determining how the terrestrial climate responds to changes in radiation input is a far from trivial problem.[4])

Indirect evidence indicates some cyclic behaviour did persist through the Maunder Minimum, but with a slightly altered period. This evidence comes from analysis of radioactive isotopes in ice cores. Energetic particles, called cosmic rays, can create radioactive isotopes in the Earth's atmosphere. The number of particles that enter the atmosphere is modulated by the influence of the Sun's extended magnetic field. At times of high activity the Sun's field prevents more particles from entering the terrestrial atmosphere because the particles find it hard to cross the lines of field. As a result, fewer radioactive nuclei are created and the concentration of isotopes in the ice cores is correspondingly reduced. These concentrations show a cyclic signature during the Maunder Minimum, with a period slightly shorter than 11 years.

There are obvious reasons for seeking an understanding of how the character of the dynamo changes during these periods, and what might drive the Sun into and out of a Maunder-Minimum-like phase.

---

[3] For an in-depth review of the topic of solar radiative output, its variability and a list of useful references, see C. Fröhlich and J.L. Lean, *Astronomy and Astrophysics Review*, 12, 2004, p. 273. Other work has looked at the Sun's modulating influence on cosmic-ray effects on terrestrial cloud cover. See, for example, N. Marsh and H. Svensmark, *Space Science Reviews*, 107, 2003, p. 317; for a general summary: M. Lockwood, in *Proceedings of the SOHO11 Symposium 'From Solar Min to Max: Half a Solar Cycle with SOHO'*, ed. A. Wilson (Noordwijk, The Netherlands: ESA Publications Division, 2002), p. 507.

[4] See D.V. Hoyt and K.H. Schatten, *The Role of the Sun in Climate Change* (Oxford: Oxford University Press, 1997); S. Solanki, *Astronomy and Geophysics*, 43, 2002, p. 509.

The nature of the various phenomena associated with activity and the cycle – spots, active regions, flares, coronal mass ejections, the solar wind – belie a crucial linking factor: magnetic fields. Sunspots are the cool footprints that mark where tubes of magnetic field pierce the photosphere. Flares are believed to result from a sudden, drastic restructuring of huge loops of magnetic field overlying the surface which cannot get out of each other's way.

Magnetic field is the common backbone. Fields organise themselves into complicated structures covering a large range of length scales in the solar atmosphere. The basic arrangement changes systematically with the 11-year activity cycle.

At the low point of the cycle weak magnetic fields pepper the solar surface in a fairly uniform manner. There is little, if any, evidence of strong concentrations of field – regions where many ropes, or tubes, of magnetic field penetrate the surface in closely packed groups, giving strong fields and high levels of magnetic flux. The large-scale field of the Sun instead resembles at this time a bar magnet with its north and south poles aligned closely to, though not exactly parallel with, the rotation axis. This gives a simple, large-scale dipole field. The lines of magnetic field are wrapped in a north–south manner around the Sun, giving what is called a *poloidal* configuration.

From this simple configuration the character of the field starts to take on a new appearance as time advances. Active regions, containing sunspots, begin to appear on the Sun. They are very choosy about where they appear – initially, centred on bands roughly thirty-five degrees north and south of the solar equator. More and more of these regions pop up. As the cycle progresses they begin forming at progressively lower latitudes. After about 5 to 6 years the number of spots and regions of strong flux reaches a maximum, by which time the features are confined in bands of latitudes roughly twenty degrees below and above the solar equator. It is now possible to see clearly that these dominant structures are no longer arranged in a north–south manner – instead lines of magnetic field are wrapped in a new, east–west *toroidal* configuration (Plate 3).

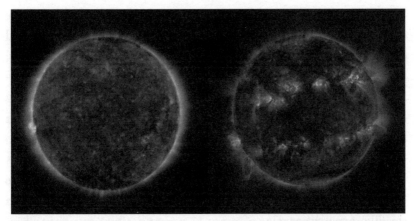

Plate 3. The Sun as observed at ultra-violet wavelengths, by the Extreme Ultraviolet Imaging Telescope (EIT) on board the SOHO spacecraft. The images show emission by iron atoms in the solar corona at temperatures of over one million degrees centigrade. The left-hand image was taken during a period of low solar activity, the right-hand image when the activity had begun its rise towards the peak level of the cycle. In the right-hand image, bands of strong magnetic field are wrapped in an east–west, toroidal manner around the Sun, at preferred bands of latitude (see text). These bands are where sunspots are found. See the plate section in the middle of this book for a full colour version. Image courtesy of the EIT Consortium, ESA/NASA.

It is believed that magnetic field just beneath the base of the convection zone is stretched out and extruded by the pattern of rotation (of which more later). The field is squashed into a smaller and smaller cross-section. This increases the strength of the field – because the density of field lines has increased – and the field becomes strong enough to react back on the plasma. The strong field then evicts plasma, and the tubes of field become lighter than their surroundings. So tubes are potentially buoyant, and a sudden instability that alters the status quo can therefore cause them to rise. When these pieces of magnetic rope pierce the surface they give rise to sunspots.

We could think of the magnetic field wrapped around the Sun as a set of elastic bands. Stretch one of the bands at a point on its circumference – this makes a kink, which can be made bigger by pulling some more. With enough stretch the top of the kink will break the surface to form a loop of field in the atmosphere. The points where it pushes through the surface are where the Sunspots are located. One spot marks where field emerges, and the other where it submerges. The pair possess magnetic fields of opposite polarity.

The toroidal field from which the spots form is aligned predominantly in an east–west direction. We can tag the polarity of the spot in a pair that leads in the direction of the rotation. The sign of this depends on whether the pair lies in the northern or southern hemisphere.

As time moves on, bands of activity in both hemispheres continue to migrate towards the equator, and the number of active regions decreases steadily. After the passage of about 11 years we return to a state where very few active regions are present. Does this cycle repeat again? The answer is yes, but with a crucial difference.

The variation in the numbers and locations of spots, and the gross level of activity, does indeed repeat as per the preceding 11-year cycle. However, the spots change in polarity. The process that gives rise to this reversal begins when the poloidal field component changes its polarity. The bar magnet flips – what was the north magnetic pole becomes the south, and vice versa. The flip takes place at about the time the spots and active regions (given by the toroidal field) are at their most numerous. The effect of the reversal begins to show up in the active regions as the following 11-year cycle begins – that is, after the old active flux has decayed away from the preceding cycle, and the new active-region fields have begun to appear. The reversal manifests as a change in the polarities of the leading spots in each hemisphere. A more complete description of the cycle therefore encompasses the 22 years or so it takes to recover the original magnetic configuration.

The structures associated with these magnetic fields can be breath-taking and reveal an elaborate pattern of magnetism in the solar atmosphere. My use above of the idea of 'elastic bands' of field is a gross oversimplification. Magnetic field on the Sun is far more complicated. The solar physicist is left in no doubt that explaining the observed architecture is going to be no easy task.

The Sun's primordial magnetic field was swept up from the interstellar medium by the gas cloud that collapsed to form the Sun. An important question we may ask is: might this primordial field be responsible for what we see on the surface of the Sun today?

The original field will have decayed over time; it is possible to show it could survive for something like 10,000 million years. At first sight this looks hopeful – this timescale is roughly twice the age of the Sun, so some vestige of a fossil field should still exist. However, we get a dose of reality when we recall the complexity of the surface fields on the Sun. It is extremely unlikely that what we observe is an ancient relic. However, the original field could serve as a seed feeding a mechanism that regenerates and re-shapes this relic magnetism into other forms that populate the cycling Sun – a mechanism like a magnetic dynamo.[5]

The basic physical mechanism underpinning the action of the dynamo is electromagnetic induction. The same effect is harnessed by a bicycle dynamo. The cyclist generates energy by pedalling the cycle. This mechanical energy is used to spin a magnetised rotor inside a conducting coil. The motion generates a current and a changing magnetic field in the coil. The current is used to power the bicycle lamp.

It goes without saying that dynamo action in the Sun – which produces solar cycle – takes place under a very different physical regime from that encountered by the manufactured components of a human-made cycle. The conducting solar plasma is not neatly constrained within a housing, is subject to complicated rotation and flow patterns and exists at temperatures and pressures not encountered in the natural terrestrial environment.

Getting to grips with how a solar dynamo might operate has occupied theoreticians since 1919, when Sir Joseph Larmor first suggested the idea. Although scientists do not have complete answers to all aspects of

---

[5] Overviews of the solar dynamo for the non-specialist are given by S.M. Tobias, *Royal Society of London Philosophical Transactions A*, 360, 2002, p. 2741; and P. Bushby and J. Mason, *Astronomy and Geophysics*, 45, 2004, p. 7.

the problem, a large and convincing body of evidence shows that dynamos do a pretty good job of describing what is observed.

The problem for the Sun has several key steps. First, a simple poloidal field has to be turned into the toroidal field needed for the sunspots and active regions, which are wrapped around the equator. These features must migrate towards the equator as each 11-year activity cycle progresses. Second, the toroidal field must be used to make a poloidal field that has had its north and south poles reversed. And third, to allow the full 22-year magnetic cycle to be completed, new spots created in a new 11-year cycle must also have their fields reversed.

The basic ingredients of a dynamo are an existing magnetic field and a conductor that can move through it. The conducting plasma in the Sun fulfils the latter role. It can generate electric fields when it moves across lines of existing magnetic field. Variations in the existing magnetic field will also make electric fields. New electric fields give rise to electric currents in the plasma and these currents create new magnetic field. And so on. In order for the solar dynamo to operate, the motions of the plasma must generate new magnetic field more efficiently than it is lost to decay, and they must do so in such a way that the requirements of the solar cycle are met.

To achieve step one, the first models made use of the characteristics of the observed surface rotation. We recall from Chapter 2 that historic observations of sunspots had demonstrated the surface rotated more rapidly at the equator than at the poles, the rotation period varying from about 25 to 36 days. This is latitudinally dependent, or differential, rotation. The first modellers had no knowledge of the motions of plasma in the interior. But they had these surface observations and realised that by placing the dynamo at the surface, in the region of this differential rotation, they could turn poloidal into toroidal field.

To see how this works, consider a single strand of poloidal field straddling the Sun from pole to pole. Remember too how the plasma and the field are tied together. Now track the field line as the Sun rotates. After one rotation of the plasma and field at the equator, the plasma and field at the poles are already lagging behind. The clean north–south line of the original field will now have a kink in it, centred on the faster-moving equator. The field line therefore has a small component in the east–west direction – the differential rotation is starting to make toroidal field.

After further revolutions it is easy to see that continued stretching of the field would wrap it up around the equator, turning what was a poloidal field into predominantly toroidal field. In the new orientation the field also points in opposite directions in the northern and southern hemispheres, explaining the observed polarities of the spots. The stretching of field in this manner is called the 'omega' effect (because omega is the Greek letter used to describe rotation).

A change of rotation rate with depth inside the Sun can have a similar effect on magnetic field buried within the interior. Take the case where the rotation rate decreases with increasing radius – a negative gradient. Next take a north–south-oriented line of field cutting through the convection zone. Parts of the field line in the vicinity of the equator will lie in layers closer to the centre of the Sun than those at higher latitudes. The faster rate of rotation found at greater depths will therefore begin to generate a kink in the poloidal field line, and toroidal flux.

This second scenario is what other, earlier models had assumed. They relied on an outwardly decreasing rate of rotation in the interior to stretch out the field and thereby spread the dynamo process across the convection zone (a 'distributed' dynamo). The zone seemed to be a good place to locate the dynamo, given the high conductivity, and motions, of its plasma. To round out the models – and complete the full cycle – the problem was then one of how to reverse the poloidal configuration.

Unfortunately, there is no large-scale way of doing this that is analogous to the omega effect. Instead, dynamo modellers have to rely on the cumulative effect of small-scale motions in the plasma. There has been much controversy over how, and where in the outer layers of the Sun, this might be achieved. Nevertheless the omega effect ensures that, provided some way can be found to recover a reversed poloidal field, the subsequent action of the differential rotation will give rise to toroidal, spot-like fields with the required reversal of the leading-spot polarities.

The first information that helioseismology brought to bear on the solar-cycle problem came from the observations of frequency shifts. In the 1980s the Nice, Tenerife, Birmingham and Stanford groups found significant frequency shifts, thereby confirming Woodard and Noyes' discovery. However, a paper published in 1990, by Ken Libbrecht and Martin Woodard, made the most important breakthrough during this period.[6] Ken's Big Bear dataset included a substantially greater number of modes, spanning as it did a much larger range in angular degree. These data made it possible to pin down the frequency shifts very precisely.

When he investigated each frequency part of the oscillation spectrum in turn, Ken discovered that within these the higher-degree, shallower-penetrating modes always suffered a larger shift than their deeper-penetrating, lower-degree counterparts. This indicated the mechanism responsible was having a harder time affecting modes that engaged deeper-lying layers, and therefore a larger volume of the interior, in oscillation. Which ruled out a stimulus originating in the core.

The fact the shifts appeared to scale in size with the inverse of the inertia associated with the modes proved to be the smoking gun. From discussions he had had with colleagues Peter Goldreich and Pawan Kumar – both theoreticians with an interest in the solar-cycle problem – Ken knew the trend fingered the layers near the surface as being those in which the shifts had originated. A dependence of the shifts on frequency – in particular the fact the low-frequency modes underwent a very weak shift – further confirmed the conclusion.

Ken and Martin were also able to demonstrate a dependence of the shifts on solar latitude. This was to be expected of a solar-cycle effect.

---

[6] K.G. Libbrecht and M.F. Woodard, *Nature*, 345, 1990, p. 771. A splendid summary of inferences about the modes made from Ken's observations (including the solar cycle and mode excitation and damping) can be found in K.G. Libbrecht and M.F. Woodard, *Science*, 253, 1991, p. 152.

The active regions are to be found only in certain bands of latitude. If their presence were in some way linked to the shifts, the shifts should have been stronger in the active regions than in neighbouring quiet patches – precisely what the analysis bore out.

These findings have been confirmed in greater detail by subsequent observations. The precision with which the shifts of medium- to high-degree modes can be studied has been enhanced greatly by the advent of the ground-based GONG network and the MDI on the SOHO spacecraft. There is no doubt the shifts of the frequencies reflect changes associated with the solar cycle. But a decade and half on from Libbrecht and Woodard's paper we are still not sure precisely what it is that is shifting the frequencies.

For the frequencies to change, one or both of the following must be happening. First, the cavities in which the modes are trapped may change in size. Second, the speed of the waves within the cavities may alter in response to cycle-related variations of the gas properties.

Magnetic fields are an obvious candidate as a driver for the variations – the frequency shifts are largest in regions where there are strong concentrations of magnetic field. Magnetic fields can affect the modes in more than one way. They can do so directly, by the action of the magnetic Lorentz force on the plasma. This provides an additional restoring force to drive the waves, the result being an increase of frequency. Magnetic fields can also influence matters indirectly, by affecting the physical properties in the mode cavities and, as a result, the propagation of the sound waves within them. This indirect effect can act both ways, to either increase or decrease the frequencies.

Any theoretical description of the frequency shifts must at the same time satisfy another constraint – the fact the solar constant changes over the solar cycle. Magnetic fields may also be the cause of this. We know sunspots are regions of strong magnetic field. They have a dark appearance because they are cooler than their surroundings. This state of affairs arises because magnetic fields inhibit the action of convection.

Sunspots have a large cross-section on the surface. But certain magnetic features with small-cross sections – the so-called faculae – appear brighter than their surroundings. In both cases the re-direction of the outward flow of energy, brought about by the inhibiting effect of the convection, means the vertical sides of the structures beneath the surface receive extra heating. This only affects the outer rim of the spot structure, but the whole of the faculae are affected because these are so narrow.

At the peak of activity there are greater numbers of spots and faculae on the surface. The dimmer spots act to lower the luminosity, the brighter faculae to increase it. An explanation of luminosity changes based on the appearance of these magnetic features then supposes the facular contribution just outweighs the spot contribution at the peak of activity. But changes to the luminosity can also arise from an effective change in solar radius – this alters the surface area through which the Sun can radiate – or thermal changes. Various models attempt to give the observed luminosity variations in which at solar maximum the Sun is predicted to be hotter (courtesy of Jeff Kuhn) or cooler (courtesy of Henk Spruit).

Attempts to reconcile the frequency and luminosity changes by a common driver, of thermal origin, encounter major difficulties (this from work by Douglas Gough and Mike Thompson, and later Nick Balmforth, Douglas, and William Merryfield). Phil Goode and Wojtek Dziembowski have recently made headway in offering an explanation, based on the effects of the magnetic fields, which may solve same of the outstanding problems.

These are some of the issues involved. The effects impart very small changes to the internal structure of the star, which is why the problem is so difficult. Extra clues are available from observations of changes in the power and damping of the modes,[7] which also correlate well with the

---

[7] The Tenerife group was the first to uncover changes in the power of the modes, in the late 1980s. Changes in the damping, also claimed about the same time by the Tenerife group, took longer for observers to reliably pin down.

activity cycle and provide information on the complicated process of convection. Any complete explanation of the cycle will need to fit all the pieces into the puzzle.

The helioseismic results for the rotation in the convection zone that have, to date, given the most direct and by far the most useful input for the solar dynamo modellers. First, it became possible to address how the differential pattern of rotation penetrated beneath the surface and whether any substantial gradients with radius were present.

Pre-helioseismic models of rotation in the convection zone – work spearheaded by the likes of Peter Gilman and Gary Glatzmaier – gave a scenario in which the rotation rate depended largely on the perpendicular distance from the rotation axis. This yielded a pattern in which the rotation was constant on cylinders wrapped around the axis – so-called 'Taylor columns'. In such a picture, small-diameter cylinders intersect the solar surface at high latitudes, whereas larger cylinders do so at low latitudes. In order to match the differential sunspot rotation (fast at the equator, slow at the poles) material lying on the surface of a small cylinder must rotate less rapidly than plasma on a larger cylinder.

An important consequence of the rotation models was that they therefore predicted an increase of rotation rate with increasing radius, a positive radial gradient. However, the distributed dynamo constructs required a negative, not a positive, gradient throughout the convection zone to give something resembling the surface pattern of activity. This was clearly a problem. Enter helioseismology.

Tim Brown made the first observational inferences on interior rotation at different latitudes,[8] from data collected by his Fourier Tachometer instrument. He demonstrated that the surface pattern did indeed penetrate the interior. The pattern also appeared to be less prominent in modes that probed beneath the base of the convection zone. This

---

[8] T.M. Brown, *Nature*, 317, 1985, p. 591.

indicated that, at the very least, the differential rotation was not as pronounced in the deep interior as it was in the outer layers.

Tom Duvall, Jack Harvey and Martin Pomerantz confirmed the main thrust of these results from analysis of their South Pole dataset. Ken Libbrecht reached similar conclusions from his Big Bear observations. And by the late 1980s, with more extensive data to hand, it became possible to perform inversions to re-construct the rotation rate at different latitudes. The first inversion used Ken's data and was calculated by Jørgen Christensen-Dalsgaard and Jesper Schou. Other teams soon followed suit.[9] The findings were completely unexpected.

The nature of the uncovered profile was such that at each latitude the rotation rate appeared to be largely independent of depth in the convection zone. This seemed to present an even bigger headache for theorists seeking to marry models of rotation to those of the dynamo. The conflict between the negative gradient of rotation needed by dynamos located in the zone and the positive gradient predicted by the rotation models had now turned into a new problem – helioseismology was saying hardly any gradient was present at all. However, another finding in the helioseismic data offered a way out.

The results painted a picture in which the pattern of surface differential rotation persisted throughout the convection zone, but altered abruptly at its base. There, it transitioned from a latitudinally dependent structure to a simple, solid-body-like profile – where the rotation was unaffected by latitude – in the deep interior beneath. From these early data it was not possible to say how abruptly this change in rotational behaviour took place. However, with modern data (see Plate 4) we now know the answer is over a few per cent of the solar radius – a very sharp change indeed. Ed Spiegel and Jean-Paul Zahn christened the site at which the change takes place the *tachocline* – *tacho* meaning 'speed'

---

[9] The authors included Tim Brown, Alesandro Cacciani, Jørgen Christensen-Dalsgaard, Wojtek Dziembowski, Douglas Gough, Phil Goode, Slyvaine Korzennik, Ken Libbrecht, Cherilynn Morrow, Ed Rhodes, Steve Tomczyk, Mike Thompson and Roger Ulrich.

(from the Greek *takhos*), and *cline* indicating a gradient (from the Greek *klinein*, 'to lean').[10]

The tachocline is located immediately below the convection zone. The steep gradient in rotation present across the tachocline – much stronger than anything elsewhere in the outer layers – makes the tachocline the obvious site in which to locate the stretching of field by the omega effect. Locating the seat of the dynamo in the tachocline is attractive for another reason. The distributed models, in which the omega effect resided in the convection zone, suffered from the fact that magnetic field rose to the surface too quickly. This meant there was not enough time for the rotation to stretch out the field sufficiently to give the strengths required to produce sunspots and other features observed at the surface. Because the tachocline lies in the radiative zone, the plasma and field are not subject to the extra buoyancy from the convection. Fresh field made in the tachocline can therefore be stored for much longer, so it can be wound up more tightly, and made stronger, before magnetic buoyancy instability can send it on its journey up through the zone to the surface and possibly beyond.

The discovery of the tachocline has allowed dynamo theorists to nail down the location of the omega effect, enabling them to turn their attention to how the toroidal field is converted back into poloidal field to close off the cycle. There are two main schools of thought.

In one – the interface dynamo of Eugene Parker – it is the so-called 'alpha' effect that closes off the cycle. The omega and alpha effects are located in close proximity to one another. The omega effect is in the tachocline. The alpha effect does its work just above, in the lower reaches of the convection zone. As a result of the Coriolis force, the alpha effect twists toroidal field that has been sent on its way upward from the tachocline below. This idea for the alpha part, proposed originally by

---

[10] Spiegel originally used 'tachycline', but deferred to (in his words) 'the terminological sensibilities of D.O. Gough' and modified the name. See E.A. Spiegel and J.-P. Zahn, *Astronomy and Astrophysics*, 265, 1992, p. 106.

Plate 4. Three-dimensional cut-away showing the rotation rate in the solar interior, as derived from Michelson Doppler Imager (MDI) observations. The tachocline mediates the transition from differential rotation in the outer layers to near solid-body rotation below. See the plate section in the middle of this book for a full colour version and further explanation. Rotation image courtesy of P.H. Scherrer and the MDI team, and ESA/NASA.

Parker in the mid 1950s, relies on the same effect that drives cyclones in the Earth's atmosphere. This is why it is sometimes given the grander name of 'cyclonic turbulence'.

An easy way to see the effects of Coriolis force is to imagine you are watching a friend who is standing at the centre of a spinning roundabout. They throw a ball radially outwards and you watch its trajectory. The ball not only moves radially away from your friend but also swings to one side. This extra component arises because the roundabout is spinning. We ascribe the effect on the trajectory as coming from something called the Coriolis force.

A cyclone forms around a depression, a region of low pressure. Air therefore falls towards the centre of the depression. This is analogous to a ring of individuals, arranged around the edge of a roundabout, who all throw a ball toward the centre. Instead of large-scale flows of air we have projectiles. The trajectory of each ball is diverted if the roundabout is spinning, each twisting in the same way.

The Earth spins too. Air falling into a depression is deviated and given a twist. The result is a cyclone. The handedness of the rotation matters – the direction of twist is opposite in the northern and southern hemispheres.

Now let us return to the Sun. Here, as material rises through the convection zone it expands. The appropriate roundabout model is now one in which we ask our crowd of willing volunteers to congregate at the roundabout's centre. They throw their balls outwards; this mimics the expansion of our bubble of material in the convection zone. The trajectories of the balls twist; so too does the bubble of gas, and the magnetic field it carries. We could equally well have described the case of a bubble that contracts in the convective down-flow. The result would be more twist, just like the cyclone. Once again, it matters whether you are in the northern or southern hemisphere. The idea is that many, similar, little twists of field merge together – via a process called magnetic reconnection – and aggregate into a large-scale poloidal field. The 'alpha' in the name of the effect is a constant of proportionality, which fixes the size of poloidal field made in this manner. Its sign indicates the polarity, or direction, of the field.

The complete picture of this form of interface dynamo is, then, one in which toroidal field is wound up in the tachocline. When this field becomes unstable to magnetic buoyancy it rises through the convection zone. Weaker toroidal field, subjected to the effects of the cyclonic turbulence, never makes it to the surface. Instead, it is turned into poloidal field and returned to the tachocline by immense, downward-moving plumes of material near the bottom of the zone, which overshoot slightly into the radiative interior and allow their cargo of poloidal flux to be deposited ready for reprocessing. This process is called flux pumping; it has been shown by the likes of Steve Tobias and Nigel Weiss to be efficient at pulling field down to the bottom of the zone. Stronger toroidal field has enough buoyancy to avoid the fate of the weaker field, allowing it to penetrate the surface to form sunspots.

Models in which the omega and alpha effects were distributed throughout the convection zone suffered from a potentially catastrophic 'quenching' problem. Studies show that the twisting needed for the alpha effect to operate gets progressively less efficient – or is 'quenched' – as the strength of the magnetic field increases. Results from the early and mid 1990s suggested this alpha-quenching became important at much lower field strengths than had previously been thought. The interface dynamo was an attempt to get around this problem. It separated out the locations of the alpha effect (still in the convection zone), and the omega effect (in the tachocline, where the strong field is stored).

The other main model of dynamo has its alpha-like effect located at the surface. Here, field rising up through the convection zone arrives at the surface with some twist. The combination of this twist and the manner in which the spots and active regions decay away at the surface turns toroidal back into poloidal field. When this idea was originally proposed, by Horace Babcock, and Bob Leighton (the very same who discovered the 5-minute oscillations) in the 1960s, the omega effect was also located at the surface in dynamo models. Today, as we have seen, there is widespread acceptance that the omega effect lies instead in the tachocline. This has changed the character of a Babcock–Leighton dynamo, in that the omega and alpha effects end up separated by the full extent of the convection zone. The separation creates the problem that some means has to be found to carry the sources for the fresh poloidal flux – the decaying active regions – all the way down to the tachocline so they can be reprocessed. Modern versions of this flavour of dynamo, now called 'flux transport' models, rely on the north–south circulation current, called the meridional flow, to solve the problem.

This current acts to move material at the surface in both hemispheres towards the poles. There has to be some form of return flow, otherwise material would accumulate at the poles. It is assumed that the returning leg of the conveyor-belt is located deep down, near the bottom of the convection zone. The resulting pattern of circulation can therefore transport the poloidal flux as required. The timescale of the flow largely controls the cycle period in these models.

Surface measurements of the magnetic fields suggest the Sun retains some memory of what happened in previous cycles. A potentially attractive feature of flux transport models is that the meridional flow can provide such memory. During any one cycle old magnetic field from previous cycles may still be present on the lower reaches of the conveyor. It is then in a position to influence the characteristics of the new field that forms.

Putting the mechanism that makes poloidal flux at the surface also gets round potential problems relating to quenching effects. However, work done by Joanne Mason, Steve Tobias and David Hughes, at Leeds, suggests that any alpha effect operating deep down may tend to drown out the effects of one operating at the surface. Recent state-of-the-art flux transport models, created by Mausumi Dikpati, Peter Gilman and colleagues, do now include an alpha term at the bottom of the zone as well.

Many issues concerning the alpha effect need to be resolved. The various models contain many free parameters, and the choice of values can have a huge impact on the solutions given. Other variants of the effect are also being tried by dynamo modellers.

To give a flavour of the complexities, take the cyclonic effect. In the older models an outwardly decreasing rotation rate was needed for sunspots to migrate towards the equator as required. Things are not so simple now. Depending on the assumed physical conditions, it is possible to get model spots that migrate towards *or* away from the equator for positive *or* negative values of the rotation gradient. This explains why the sign of the gradient at the tachocline is not a cause for concern. You may have wondered why I had not addressed this issue – at latitudes below about thirty degrees the rotation rate decreases with increasing radius, which would not have been acceptable in the old treatments. By tuning the parameters that describe the alpha effect it is possible to obtain a pattern of spot migration that superficially matches what is seen. But this is somewhat ad hoc and the alpha effect remains poorly understood.

The existence of the tachocline has raised several fundamental questions about the Sun's dynamic evolution. This thin layer matches the transition in rotational behaviour above and below, and as the star evolves must mediate the transfer of angular momentum from the immense reservoir in the core to the outer envelope and beyond.

In order for the rotation to change its character something must be acting in the radiative interior to mix angular momentum in latitude so that the differential rotation from the convective zone above is smoothed out, or destroyed, below. But any mechanism would presumably act in the radial direction (i.e. in depth) too – and this creates a problem.

If we assume the efficiency of the mixing is the same in all directions – isotropic – we would expect the mechanism to have already done a reasonable job of removing the abrupt radial gradient of rotation at the tachocline as it has demonstrably done in the latitudinal direction. The result? The tachocline would then be much wider than is actually observed. So, from a list of suspects we need a mechanism much less effective in the radial as opposed to the latitudinal direction to explain the narrow width of the tachocline. A likely possibility is the effect of a magnetic field threading the radiative interior.

Douglas Gough and Mike McIntyre have considered circulations that penetrate from the convection zone into the tachocline and are then diverted by a weak magnetic field further down. The field acts to prevent the tachocline spreading out in radius, in effect forming a firm lower boundary. The pattern of rotation at low and high latitudes acts to keep field bottled up in the radiative interior. This field need only have a strength a tiny fraction of that at the surface in order to give the required effect. What is more, it would then also be a prime candidate for enforcing the solid-body-like rotation that, as we saw in Chapter 9, is present in the radiative zone.[11]

One final comment about this scenario. It is at latitudes of about thirty degrees that the rotation rates match in the deep interior and convection zone. Plasma and magnetic field might therefore be dragged up from the interior at these locations. These happen to be close to the latitude at which active regions first appear.

Another possible mechanism may relate to the effect of turbulent flows. These do know about the change of pressure and density with depth – the radial stratification inside the Sun – and so are aware of the difference between the latitudinal and radial directions. It may therefore be possible to envisage this turbulence acting efficiently in latitude but being inhibited by the change of properties with depth.

With a new picture of the interior rotation beginning to form, the next step was to look for variations in the rotation over time, especially variations well correlated with the solar activity cycle. Changes in the differential rotation profile on the solar surface had been uncovered as early as 1980 from observations that tracked features across the photosphere. These data, collected by Bob Howard and Barry LaBonte, showed bands of plasma at particular latitudes rotating either slightly faster or slower, by a few per cent, than the level expected from the smooth, underlying pattern of rotation. What is more, the bands shifted position as the solar cycle progressed, tracking towards the equator on a timescale that suggested they carried the signature of the effects of the cycle. These flows were called torsional oscillations.

Ken Libbrecht and Bob Woodard found in the Big Bear data the first indications of changes beneath the surface. Then came analyses by Jesper Schou and Sasha Kosovichev of f mode data, which showed the torsional

---

[11] The fact a weak interior magnetic field can enforce solid-body-like, rigid rotation was known long before this. However, Gough and McIntyre's paper offered a unified explanation of how magnetic fields can influence the structure of the deep radiative interior and tachocline.

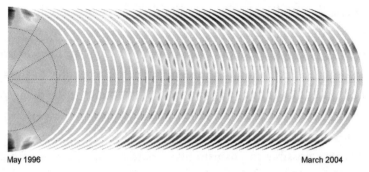

May 1996   March 2004

Plate 5. The 'torsional' oscillations in the outer layers of the solar interior, as revealed from the analysis of 72-day chunks of data collected by the Michelson Doppler Imager (MDI). The large-scale pattern of differential rotation seen at low activity has been removed from each set, leaving the much weaker torsional signal. Over time, bands at moderate latitudes migrate towards the equator (lighter patches close to the equator), while a strong poleward-moving flow is present at high latitudes. See the plate section in the middle of this book for a full colour version and key. Image courtesy of S. Vorontsov and collaborators.

oscillations at the surface. By the late 1990s sufficient quantities of medium- and high-degree data had been collected – courtesy of the MDI and the ground-based GONG network – to probe the extent to which the torsional oscillations penetrated the interior. The interior flows uncovered by helioseismology (Plate 5) nicely followed their surface counterparts.[12] What was new was that it was now possible to see that these extended in depth throughout a good fraction of the convection zone. Furthermore, a previously unseen pattern of flow, which migrated towards the poles, was found.

That these interior-penetrating, torsional oscillations are conspicuously present in the data is beyond doubt. However, the nature of the p modes from which the flows are extracted may be fooling us into thinking the bands are present right around the Sun. The input data are

---

[12] See R. Howe, J. Christensen-Dalsgaard, F. Hill, R.W. Komm, R.M. Larsen, J. Schou, M.J. Thompson and J. Toomre, *Astrophysical Journal*, 533, 2000, p. L163; and S.V. Vorontsov, J. Christensen-Dalsgaard, J. Schou, V.N. Strakhov and M.J. Thompson, *Science*, 296, 2002, p. 101.

rotational splittings from global modes formed by waves that live long enough to make a complete circuit of the Sun. Global modes provide a measure of properties *averaged* in solar longitude. Although the global modes allow differences in structure to be inferred in latitude, it is not possible to discriminate the structure north and south of the equator. Only those features that have structure that is *symmetric* about the equator can be uncovered from the global-mode data.

The observed behaviour of the flows is certainly suggestive of a signature of magnetic effects from the solar cycle. An obvious candidate is the back reaction of the magnetic field on the solar plasma (via the magnetic, Lorentz force). Dynamo modeller Steve Tobias notes that the fact the flows penetrate through the zone is precisely what one would expect. The force acting ought to be largest in the vicinity of the tachocline, where the fields are strongest. But the density of material is high at the tachocline, so the effect on the rotation will be smaller than at the surface and therefore harder to see.

It is possible to study effects like this in dynamo models by including the back reaction from the magnetic field. The models are then called 'dynamic'. The uncovered poleward flow is harder to understand; it may suggest the presence of substantial field at high latitudes. If the magnetic explanation is along the right track, then one might anticipate finding the torsional bands only in regions that harbour strong magnetic fields. Clearly, a finer tool is required to target particular patches and probe the volumes beneath them. Helioseismology has an answer to this problem, as will be revealed later in the chapter.

In 2000, further signatures of changes to the rotation were found in GONG and MDI data. These were even more intriguing and certainly more controversial. Rachel Howe and collaborators uncovered evidence for variations at the base of the convective envelope.[13]

---

[13] R. Howe, J. Christensen-Dalsgaard, F. Hill, R.W. Komm, R.M. Larson, J. Schou, M.J. Thompson and J. Toomre, *Science*, 287, 2000, p. 2456.

The changes appeared to be most prominent in the low-latitude regions, just above the base of the convection zone. At the same time there were suggestions of variations in the complete opposite sense – in anti-phase – some 60,000 kilometres deeper down in the outer parts of the radiative zone. The sizes of these changes seemed to imply the rotation was speeding up, then slowing down – or vice versa, depending which side of the tachocline one was probing – by roughly one per cent on a period of about 1.3 years. The variations uncovered by the analysis then all but disappeared when mid-latitude regions were tested. A periodic-looking signal closer to 1 year was found when attention was focused on the plasma sixty degrees north and south of the solar equator.

The appearance of a 1-year signal made the collaborators a little nervous, since this is a timescale that leads the finely tuned senses of the helioseismologist to smell an artefact. As the GONG data come from a ground-based network, they can be subject to subtle seasonal variations. Furthermore, the GONG instruments and the MDI, the latter tucked on board the SOHO spacecraft, both orbit the Sun with a period of 1 year. Over this time both vary their distance from and velocity with respect to the Sun, so yearly effects need to be properly dealt with in the preparation and calibration of the raw data. Changes having a period closer to 1.3 years are much harder to explain.

Is this question of timescale the main origin of the controversy? Not really. The greater issue has been the level of significance ascribed to the supposed variations. Sarbani Basu and H. Antia question whether the 1.3-year signal really is solar in origin. They have made their own analyses of the same data and have failed to uncover a similar signature. Moreover, the variations found by Rachel and her colleagues appear to come and go with time. Over the stretch from 1996 to 2000 they were visible. They then disappeared for two years or so, before reappearing in more recent data. This may or may not be a problem, depending on how one looks at things. Who is to say the effect, if real, is not one that is transient in nature?

The solar wind shows evidence for periodicities close to 1.3 years. Auroral displays in the Earth's atmosphere – the Northern and Southern Lights produced by energetic solar wind particles interacting high up – do too. There are even suggestions of similar behaviour in sunspot observations. This is good news for those who feel the signal is real. Similar effects may also be present in other solar measures that relate to activity. What is more, these effects appear to come and go, just like the apparent variations in the rotation either side of the tachocline.

The analysis needed to uncover the variations is complicated. Furthermore, no two people will conduct this in exactly the same way. Such is the kind of debate that arises when an exciting and new but marginal and controversial result is uncovered.

The fact that some analyses show the 1.3-year signal whereas others do not is clearly a cause for concern. Several things are needed to resolve the issues. Continued collection of data will allow fresh analyses of the signal to be made. Furthermore, extending the analysis to data other than GONG's and the MDI's is desirable. The ECHO data of Steve Tomczyk are being scrutinised. What may also help to resolve the question of the come-and-go nature of the signal is the possible use of the Mount Wilson data of Ed Rhodes. Ed and his team have been painstakingly reformatting data stretching back to 1988, a good 8 years or so before GONG and the MDI began operating. Jesper Schou has already started looking at the torsional oscillations; the deeper 1.3-year signal possibly lies in wait.

Our debate here has revolved largely around the issue of whether the variations are real. Let us suppose for the moment that they are. What, then, might they be the signatures of? Several possibilities have been raised. The fact that the signals uncovered above and below the tachocline are in anti-phase – as one region speeds up, the other slows down – suggests some type of sloshing back and forth of angular momentum.

Dynamo models studied by Eurico Covas and colleagues show some evidence for chaotic behaviour and do so in such a way as to enter new states with periodicities that are sub-multiples of the 11-year solar cycle. One of these corresponds to 11 divided by 8 years, not too far from 1.3 years.

Mike Thompson, who is one of Howe's team, points out that the phenomenon need not necessarily have its origin in the tachocline region. The apparent presence of the effect in the solar wind does suggest a magnetic link of some kind. This might be the result of a modulation in the structure of the magnetic field deep down, near the tachocline, which then moves up through the convection zone to eventually affect the solar wind; or, it might originate near the surface and then spread *up* into the wind and *down* into the interior. Such cause and effect should bring phase shifts – delays in time between the occurrence of the change in structure and the appearance of the effect in regions where it is observed. As more data on the phenomenon are collected we will be in a better position to answer these complicated questions.

All of the analyses referred to above, and indeed mentioned throughout the book so far, have involved the use of global modes. These modes provide a measure of properties that averaged right around the Sun (in longitude) and are symmetric around the equator (in latitude). This presents a disadvantage if one wishes to sample the physical conditions beneath particular small patches of the surface. We have already referred to this in the context of the torsional oscillations. Other features relating to the solar cycle cry out for a similar, scalpel-like tool. For example, what is the structure like beneath sunspots and active regions?

This is not to say such information is absent from the global modes and could not, at least in principle, be extracted. The information is there, but because very many modes are affected by any localised feature – and lots of acoustically interesting features are present at any one time – disentangling the information would be a monumental task. However, a practical option is available – local helioseismology. This class of techniques look at what sound waves are doing 'locally' beneath small patches of the surface, rather than waiting for some average pattern of behaviour to establish globally. These techniques are therefore ideally suited to solar cycle studies.

The local helioseismologist's toolkit of techniques was stocked initially with two methods: ring diagram analysis and time–distance analysis.

Frank Hill developed ring diagram analysis in the 1980s. He was drawn into helioseismology as soon as he started graduate school, at the University of Colorado, in September 1976. Like Ed Rhodes, his sights were set initially on the physics of plasmas. One of the first lectures Frank attended was a mathematics class given by Juri Toomre. While chatting after the lecture, Toomre brought up the fledgling topic of helioseismology. Juri had a particular interest in the solar oscillations because they offered a potentially exciting means of probing the physics and dynamics of the convection zone (a line of study that Juri remains in the forefront of today). Frank was asked if he might be interested in getting involved. To use his words, he 'got sucked in' by the intriguing nature of the proposal.

Frank worked for his graduate thesis on using oscillations as a probe of convection in the Sun. Some of the data he needed were collected at the Sacramento Peak Observatory. The idea for what would become the ring diagram technique came out of brainstorming sessions at the observatory, in the early 1980s, with another of Juri's students, Larry November. Frank and Larry would sit around at night, talking over various physics problems. During one of these sessions they started to think about how the traditional k–ω diagram might be turned from a two- to a three-dimensional power spectrum. Their starting point was the fact that the sound waves, once generated, travel in all directions. Might it be possible, they wondered, to track and measure the properties of these waves in all three dimensions? They soon realised the answer was a resounding yes.

The basic k–ω diagram provides information on how the frequencies of the modes formed by the trapped waves depend upon horizontal wavenumber or wavelength. One might choose to discriminate the wave properties in different directions on the surface. This can be achieved by resolving the horizontal wavenumber into directions, say, parallel to the solar equator (call that x) on the one hand, and parallel to the rotation axis (y) on the other. One now has three pieces of information – frequency, a wavenumber in x and a wavenumber in y – where before we had just two (frequency and a single resultant wavenumber). This information can be rendered on a three-dimensional plot, with frequency filling the vertical axis. The form the new diagram takes is tantamount to rotating the traditional k–ω plot around the frequency axis (Figure 10.2a). Each ridge now forms a three-dimensional surface shaped just like the open bell of a trumpet.

The outermost trumpet is formed by the f modes (the surface gravity waves). These modes formed the lowest ridge on the two-dimensional plot. Successive overtones of the p modes then form trumpets nested within the f mode trumpet. Each trumpet gets successively smaller in size, giving something akin to a helioseismic set of Russian dolls. Now take a slice through the nested trumpets at a constant frequency. We will be left with a plot containing rings each marking the location of its trumpet surface at that frequency (Figure 10.2b).

What can we learn from these rings and trumpets? A powerful application of the method is to investigate bulk flows of plasma beneath the Sun's surface. This can be done from observations of small patches of the surface, building a picture, piece by piece, of the underlying currents of the rivers of solar plasma.

Bulk flows leave a characteristic imprint on the rings. Suppose that beneath a chosen patch of observation there exists a strong current of flow in a particular direction. This additional 'river current' will help to carry the wave fronts of the sound waves in the same direction. In scientific parlance, the waves are 'advected' so that those moving with the flow have their frequencies increased; those moving against the flow have theirs decreased. Waves that move at right angles to the flow are unaffected.

(a)

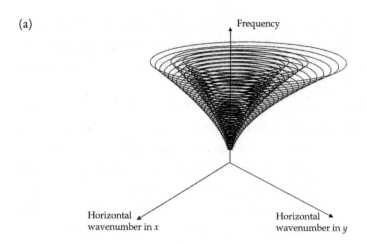

Frequency

Horizontal
wavenumber in x

Horizontal
wavenumber in y

(b)

Cut at 3.5 milli-Hertz

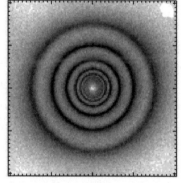

Horizontal wavenumber in y

Horizontal wavenumber in x

Figure 10.2. (a) A three-dimensional power spectrum, with horizontal wavenumber axes and a vertical frequency axis. Ridges from the traditional two-dimensional k–ω diagram are now nested, flaring, three-dimensional surfaces called trumpets. Source: Hill, *Astrophysical Journal*, 333, 1988, p. 996; reproduced with the kind permission of the American Astronomical Society.
(b) When a slice is taken, at constant frequency, through the three-dimensional trumpet surfaces, a two-dimensional spectrum consisting of rings is obtained. Image courtesy of D.A. Haber, B.W. Hindman, R.S. Bogart, M.J. Thompson and J. Toomre, from analysis of MDI data.

The combined effect results in the trumpets tilting over. The rings are then not only shifted slightly but also distorted from a circular shape. By measuring this change in geometry it is possible to infer the direction and size of the acting flow.

Some care is needed in the analysis. The frequency and wavenumber of a mode fix the range of depths it probes. By taking slices through the trumpets at different frequencies and then looking at the shift and distortion of the resulting rings, it is possible to say something about how any bulk flow varies with depth (in both magnitude and direction). Changes in temperature and the magnetic fields can also leave their imprint on the rings. Variations in the thermal properties of the layer can increase or decrease the diameter of the rings; variations in magnetic field will, like flows, distort the rings because the effect they have is direction dependent.

Cutting down the size of patch observed brings the tantalising prospect of being able to study the effects on the wave properties of individual active regions. One might think the smallest size possible would be set only by the spatial resolution of the detector – i.e. the number of pixels the visible disc is split into. However, the uncertainty principle has something to say on this matter. The smaller the observing patch, the poorer is the resolution in wavenumber. This means a compromise has to be struck. The practical lower limit is a patch whose sides are about 200,000 kilometres long.

Frank and Larry's initial discussions allowed them to work through the concept of rings and trumpets. Other commitments then intervened. Larry had a first stab at making a three-dimensional power spectrum a couple of years later, but took the work no further. In the end, what galvanised Frank into putting things down formally on paper was a bit of old-fashioned scientific competitiveness.

He found out that Tim Brown and Cherilynn Morrow were about to write a paper on the rings and trumpets concept. Not wanting to be beaten to the punch, Frank set to work. The resulting paper, published in 1988, marked the introduction of the technique to helioseismology.[14] It laid out the principles of ring analysis and also contained some analyses of real data. Although the concepts were sound and robust, Frank soon realised there were potentially better ways to develop the analysis recipes. Together with Jésus Patrón, who came over from the Instituto de Astrofísica de Canarias, Frank tried a new approach, and this forms the basis today of how ring analysis is performed. Juri Toomre, Deborah Haber and Brad Hindman deserve mention for having further developed the technique to the point where it is used widely as a powerful diagnostic tool.

The second staple technique of the local helioseismologist – time–distance analysis – derives from terrestrial seismology.

Geoseismologists are able uncover the nature of the structure beneath their feet by making use of both natural and human-made earthquakes. Most of the Earth's interior is solid rock, which means that energy from a seismic disturbance at the surface – be that from tectonic plates slipping past one another, or from a well-directed explosion of dynamite – gives rise to two sorts of interior waves: primary (P) and secondary (S).[15] Primary waves are compressional waves like the acoustic waves inside the Sun. Secondary waves are shear waves, akin to those on a string, and move material up and down at right angles to the direction along which the wave travels. This shears rock as the waves pass through.

Both types of wave are refracted by an increase in density as they travel into the Earth's interior. Well-placed receivers can then be used to diagnose the interior structure by analysis of, for example, the time taken for injected waves to reach the detectors from the seismic source. This analysis gives a set of time–distance plots.

Primary waves take their name because they arrive back at the surface before secondary waves launched from the same location (the primary

---

[14] F. Hill, *Astrophysical Journal*, 333, 1988, p. 996.

[15] There are also two types of surface wave: Love waves and Rayleigh waves, the latter analogous to waves on the surface of water or the buoyancy waves on the surface of the Sun (which give the f modes).

waves travel faster). Measurement of the difference in the P and S wave travel times at three well-separated detectors can be used to pinpoint the precise locations of earthquakes by triangulation of the recorded signals.

Aside from the interior stratification – the steady increase of density, pressure and temperature with depth – the delineation of the Sun into different structural layers leaves a characteristic acoustic imprint. For example, the sound speed exhibits a kink at the boundary between the convection zone and radiative interior. Waves can suffer phase shifts as they pass through this region. These signatures are fairly subtle. Those within the Earth can be harsher and more conspicuous.

Delineation of structure can mark transitions between different rock types. The course of waves can alter quite markedly as they travel across these boundaries, and reflections can also result. Andrija Mohorovičić uncovered the first major boundary. He recorded travel-time information sufficiently far from an earthquake that the detected waves penetrated deeply enough to cross the crust–mantle boundary (now commonly referred to as the 'Moho', in honour of its discoverer). Because the waves had travelled beneath the crust to a layer of higher-density rock the average wave speed and total travel time for the journey were altered. Mohorovičić inferred the properties of this new layer from the travel-time information.

One of the layers deep within the Earth is not rock. The outer core is composed of molten metal. Shear waves cannot propagate in liquids or gases. Compressional waves have no such difficulty. Primary waves injected at a steep enough angle will happily pass into this liquid layer, have their trajectory further altered by the effective change in properties they experience, and then eventually reappear on the other side of the Earth. Since the liquid core cannot support the shear waves, there is a large 'shadow' region on the other side across which S wave signals are absent.

Given the earlier discussion about the solar dynamo, it is worth adding that this liquid core is believed to be the seat of the Earth's dynamo. The Earth has a much simpler magnetic field than the

Sun – a dipole (bar-magnet) type field. The presence of a dynamo, to regenerate this field, is inferred from the time it would have taken a primordial dipole field to decay, somewhere on the order of a paltry 10,000 years. Just as on the Sun, this magnetic field reverses its polarity. However, the geological record suggests it does so in a rather intermittent manner, the time between reversals ranging from a few tens of thousands to a few million years.

Time–distance helioseismology is the solar analogue of the geoseismic time–distance method. The analysis required is, however, somewhat trickier. Geoseismologists deal with seismic waves emanating from individual sources. They may even actively experiment by injecting their own waves below the surface of the Earth. Interpretation of travel-time data recorded by suitably positioned receivers is a reasonably straightforward proposition. Contrast this with the Sun. We cannot actively experiment on the Sun (at least not at present). We must therefore make do with its rich, intrinsic spectrum of acoustic waves. But there is an obvious complication. We know acoustic waves are created by turbulence at the top of the convection zone, but this does not happen at just one location on the surface. Rather, there are many sites peppered across the entire Sun where the waves are produced. We must therefore envisage a situation in which not one, but many, sources inject waves beneath the solar surface. The problem is then one of disentangling this information to recover meaningful travel-time information. The scientist who worked out how to do this is Tom Duvall.

To understand the procedure, let us begin by zeroing in on one of the many sources of waves at the surface of the Sun. The source injects waves beneath the surface. The more nearly vertical a wave, the more deeply it will penetrate the interior and the larger will be the horizontal distance traversed before the wave reappears at the surface.

Let us assume waves get reflected when they return to the surface. Provided the pattern of interior conditions is the same beneath different patches each wave will reach the same depth, and travel the same surface distance, on its next 'skip' through the interior. The wave speed along each trajectory fixes the time taken to make one skip. A journey containing two skips will take twice as long as a journey containing one (ignoring some minor corrections we shall not worry about here). From measurements of the time taken for waves launched from the source to reappear at known distances on the surface, it is therefore possible to map the interior wave speed with position.

But how does one cope with many sources? Tom Duvall was pondering this very problem during the summer of 1992. He was on a two-week driving holiday that would take him from Tucson to Vancouver. On the first or second night (Tom is not sure which) he was lying in a motel bed when the answer struck him. Correlation of signals was the key. So it was that Tom has his second eureka moment (the Duvall law being the first). The holiday with his girl friend had only just begun and so he had to wait two weeks to try out the idea.[16]

Tom had realised he could use obvious similarities between wave signals at different points on the surface. Let us take two such locations. The separation between them corresponds to the single-skip distance *only* for waves launched at one particular angle. The signals at these locations are strongly correlated at a time difference that equals the travel time for this skip. That is, if we record the signals at the two points at multiples of the single-skip time, they will track one another in a non-random fashion. If, on the other hand, we choose some time in between, the correlation will break down.

So, for a given separation on the surface there is a special value of the time lag at which the correlation will be strong – the 'travel time' for that separation. By searching for the best correlation time, over a whole host of different separations, it then becomes possible to construct a solar analogue of the time–distance plot. In practice one uses many locations on the surface to form the correlations. The resulting plots also show evidence of correlations arising from waves that have made more than one

[16] T.L. Duvall, Jr, S.M. Jefferies, J.W. Harvey and M.A. Pomerantz, *Nature*, 362, 1993, p. 430.

skip. Single-skip correlations give the strongest time–distance curve; the multiple-skip curves are less prominent.

When applying this technique, helioseismologists are careful to talk in terms of wave, not sound, speed. This is because they are probing directly the former – the latter may be the biggest contributor to the speed at which the wave moves, but the likes of magnetic fields can affect the speed too.

The time–distance method provides the helioseismologist with a much finer tool than the ring method; it makes it possible to analyse patches on the surface as small as 3000 kilometres. The method continues to be developed.

The main use that ring analysis has been put to is to measure flows beneath the surface, superimposed on the large-scale rotational behaviour. The patterns revealed in the outer 10,000 kilometres or so of the convection zone resemble maps of terrestrial weather, which show the flow of wind in the Earth's atmosphere marking out regions of low and high pressure. This is why we talk of the solar plots (like those in Plate 6) as showing solar sub-surface weather.

On the Sun, the flows are those of the plasma, and have velocities up to about 100 kilometres per hour when viewed on the 200,000-kilometre-square scale of the ring analysis – similar terrestrial wind speeds would be equivalent to a force-ten gale. These speeds are about one-and-a-half per cent of the size of the typical rotational velocity at the solar surface. The flows are diverted and disrupted in the vicinity of strong magnetic fields – the same regions marked by sunspots and other activity.

By using modes that probe different depths in the near-surface layers, it is possible to build a picture of not only how flows vary with depth, but also (with some clever analysis) the size of the vertical components of the flow. Understanding the nature of the uncovered flows, and their variations over the solar cycle, is a major goal in helioseismology.

Time–distance analysis has confirmed the ring analysis results. Because time–distance can probe structure on much finer length scales than ring analysis, the flows are revealed to be much stronger in size. When averaged to the larger length scales commensurate with the ring analysis the flows are found to agree nicely. This provides some measure of reassurance for both techniques.

In addition to studying zonal flows, which act in the same east–west sense as the rotation, it is also possible to reveal the north–south component, the so-called meridional flow. As we have seen, this acts to circulate material in huge, conveyor-belt-like patterns in the northern and southern hemispheres. Early measurements of the surface leg of the circulation indicated the presence of a single conveyor (or cell) in each hemisphere. However, as the level of magnetic activity on the Sun rose in 1999, the northern conveyor began to show unexpected behaviour (as revealed in work by Deborah Haber and her colleagues[17]). While the poleward flow persisted close to the surface, those regions deeper than about 5000 kilometres, at latitudes higher than roughly forty degrees north, showed evidence of a flow oppositely directed towards the equator. This finding has raised important questions not only for those seeking to model the sub-surface flows but also for advocates of flux-transport dynamos, which rely on the meriodional circulation to return magnetic field to the tachocline. Mausumi Dikpati and colleagues have begun to look at the effects of multiple cells on this class of dynamo.

Perhaps the most striking results from time–distance analysis have been the reconstruction of the wave speed beneath sunspots, for example from work done by Sasha Kosovichev and his colleagues at Stanford (Plate 7). In the region immediately beneath spots the speed is found to be reduced compared with that in the surrounding magnetically quiet areas. However, at depths greater than a few thousand kilometres the reverse is true. The horizontal flows also have interesting patterns in the vicinity of the spots. These flows carry material in towards the spot in the near-surface layers, while strong downward flows surround spots. Deeper down, a ring of up-flowing material is found.

---

[17] Brad Hindman, Juri Toomre, Rick Bogart, Rasmus Munk Larsen and Frank Hill.

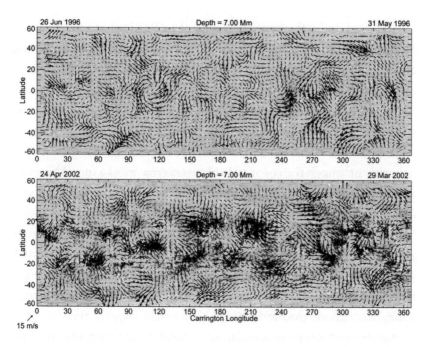

Plate 6. Plots of the solar sub-surface weather inferred from local helioseismic techniques (arrows). These are shown for a depth 7000 kilometres below the solar surface. As the Sun rotates, different parts of the surface become accessible for observation. After about one month most of the near-surface material will have made a full rotation; analysis of these data allows a map to be made of the flows found over the surface during this time (as shown above, with regions towards the right from observations made at later times, when the rotation swept them into view). See the plate section in the middle of this book for a full colour version. Image courtesy of D.A. Haber, B.W. Hindman, R.S. Bogart, M.J. Thompson and J. Toomre, from analysis of MDI data.

We know sunspots are cool because the magnetic fields they contain inhibit the upward flow of energy by convection. The bottom of the spot structure, within the interior, acts like a plug and heat builds up beneath. A plausible picture is then one in which warm material is pushed upwards, around the sides, from the bottom to meet material that has sunk around the spot from the top.

Observations of this type are allowing us a first look at how sunspots modify their surroundings beneath the surface. It is only by fine-tuning of the local techniques that we will be able to uncover the true nature of the spots. At present we do not know whether they are monolithic structures – huge tubes of magnetic flux – or fibril structures made up of several thinner magnetic ropes that are intertwined.

A much desired result would be to actually observe ropes of magnetic field rising through the convection zone on their way to form the spots and other features we see at the surface and above. There is in principle plenty of signal amenable to seismic observation. But the problem is one of (as Tom Duvall puts it) 'harnessing it in the right way', in particular to get around what is called 'realisation' noise, the intrinsic noise imprinted on the observed acoustic waves by the turbulence that excites them. New techniques will probably be needed. Tom or one of his colleagues needs to have another eureka moment!

Results of this type would also allow us to find out what happens to the magnetic field that does not reach the surface. Only some field is assumed to make the journey successfully. In simulations it can be hard to make the most buoyant fields break the surface. This is because they slowly lose their buoyancy as they rise through the convection zone. At the bottom they are notably less dense than their surroundings, but when they reach the top the deficit is greatly reduced, so they lose much of their upward acceleration. In models and simulations, adding some twist to the magnetic ropes can circumvent this problem. The extra spring helps the field rise into the atmosphere.

With the right techniques and the right data we may then be able to see whether or not a layer of flux is trapped in the convection zone. Flux like this, fairly close to the surface, would presumably play some role in flux-transport-like dynamos. Deeper down we know that twists in the magnetic field, and the way this field might then be 'shredded', are important to how the alpha effect works in interface dynamos. Helioseismology has fixed the location of the omega effect and should have an important part to play in tying down fully the alpha effect.

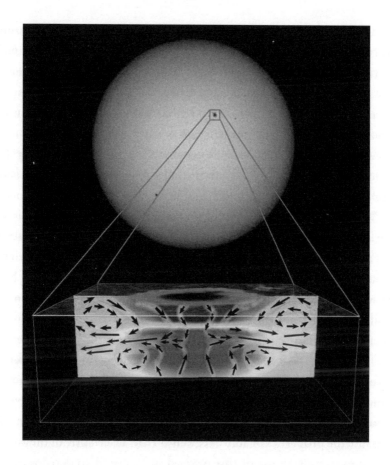

Plate 7. Variation of wave speed and flow of material (arrows) beneath a sunspot. See the plate section in the middle of this book for a full colour version. Courtesy A.G. Kosovichev, the MDI team, ESA/NASA.

Other Earth-bound local methods are beginning to be applied to the Sun. In addition, from the early 1990s Doug Braun and Charlie Lindsey began to develop holographic techniques that allow features like active regions to be imaged on the far side of the Sun.[18]

[18] See C. Lindsey and D.C. Braun, *Science*, 287, 2000, p. 1799.

Take two small, nearby source regions both of which generate sound waves. These acoustic sources are buried beneath the surface and emit waves in all directions. Waves that travel upwards will interfere and give rise to some pattern of peaks and troughs on the surface. Braun and Lindsey's technique starts from this surface pattern – the observed 'wave field' – and runs through the process in time reverse. Complicated mathematical calculations seek to re-construct the sources that gave the observed field. This allows acoustic images to be made at different focal planes in the interior.

When the plane is at a level that contains one or other of the sources, the features will appear in focus in the acoustic image; when the plane is misaligned, source images appear instead as defocused patches. This clever process can also work for features on the other side of the Sun, because waves that originate there can contribute to an observed wave field on the near side, thanks to their skipping and bouncing round the interior.

The fact magnetic regions act to absorb sound waves was established in the late 1980s by Doug Braun, Tom Duvall and Barry LaBonte. Active regions can delay the sound waves and so leave their imprint on the wave field observed on the near side. It is then possible to infer, locate and measure the properties of active regions on the unseen side of the Sun.

An exciting spin-off of this method is its use as an early-warning alarm of active regions that may end up throwing vast quantities of highly energetic particles in our direction. A regularly updated activity map of the far side of the Sun is provided on the Stanford-based MDI website. Helioseismology is therefore in a position to provide warning of the likes of the 1989 event that disrupted the Quebec power grid.

11

THE PRESENT AND FUTURE

Helioseismology has entered its fourth decade in a healthy, vigorous and robust state. As it has developed, it has also diversified. This has given birth to sub-fields – the division into global and local helioseismology springs immediately to mind.

The picture that global helioseismology has given us of the Sun's interior has been, in many respects, a reassuring one for stellar evolution theory. Our understanding of how an embryonic Sun-like star grows up to become a Main Sequence star, and then continues its evolution, remains largely unaltered. Helioseismology has enabled scientists to refine the models to a new level of sophistication. This has all been made possible by the precision and accuracy with which the Sun's interior can be probed. The fact the frequencies of the resonant modes can be measured to precisions as high as a few parts in a million makes it possible to ponder questions of the interior that previously there was no point in posing.

A good case in point is the gravitational settling of helium and heavy elements in stars. As we have seen, this had a long history before helioseismology – but it was only after it became possible to probe the interior conditions with great precision that settling and diffusion in

Sun-like stars became potentially measurable. Settling and diffusion are now included in the standard models.

This is also one of several conspicuous examples where an inference made by helioseismology has had implications for another area of astrophysics – here, the fixing of a lower-limit estimate on the age of the universe from determination of the ages of stars in globular clusters. Helioseismology has enabled scientists to treat the Sun as a vast astrophysical laboratory in which the physics of matter can be tested under exotic conditions. Inference on the opacity of stellar material is a prime example. Here, the seismic data helped to drive improvements in the complicated calculations needed to describe the multitudinous interactions taking place, once more reaping benefit for other areas of astrophysics, and leading to significant improvements to the models. Fractional changes to the model sound speeds given by tweaks in the opacities were at a level of at most one per cent, and typically less than this throughout most of the interior. Again, because the helioseismic data were so precise these differences stood out as significant discrepancies. Helioseismologists have also used the Sun as an equation-of-state laboratory, testing the thermodynamic properties of the stellar gases; the modifications these led to have again had implications for the ages of stars in globular clusters.

Improvements to our knowledge of what is going on in the interior have come on leaps and bounds. Settling is an example. So is knowledge of some of the fundamental data needed to make a solar model, such as the helium abundance. The latter's value was up for grabs in the 1960s and 1970s. As we have seen, solar models with a fractional abundance by mass of less than twenty per cent were for a time in fashion as a possible solution to the solar neutrino problem. Although no one has yet been able to fashion a direct measure of the helium abundance from helioseismology – independent of the calibration of solar models – it remains possible to rule out a low abundance because of the discrepant sound-speed profile that is obtained. Note that limitations on attempts to obtain a direct measure are not due to the quality of the helioseismic data. Rather, they arise from shortcomings in the description of the physics of dense plasmas.

An impressive collaboration – comprising Douglas Gough, Sasha Kosovichev, Jørgen-Christensen Dalsgaard, Mike Thompson and Wojtek Dziembowski – sought to obtain a direct value for the helium abundance and were able to find a pure thermodynamic quantity that depended on the effects of the ionisation of helium in the outer layers of the Sun, and therefore on the amount of helium present. This thermodynamic quantity could be measured with the helioseismic data. However, in the mid 1990s the group found, much to its disappointment, that the inferred amount of helium depended on position in the convection zone. Since the zone is always well mixed the result made no physical sense. Something was clearly amiss; it turned out that the dense-plasma physics was not good enough to describe the ionisation in the convection zone. Recent work suggests the problem still remains.

Helioseismology has subjected to the sternest interrogations models possessing a variety of additions, creases and tweaks that were hoped might crack the solar neutrino problem. Every model that seemed to solve the solar neutrino problem was found wanting. From the early 1980s onwards the helioseismic results pointed consistently to a remedy whose origin lay in the physics of the neutrinos. By the beginning of the 1990s this pointing had became a vigorous wave accompanied by a loud holler, which only grew louder still as the decade progressed. The same thing was said all along – and with some confidence. The role the oscillations have played in solving the solar neutrino problem represents one of the most important triumphs of helioseismology, one whose significance is often overlooked beyond this field.

Rapid rotation of the deep interior also offered one possible way out of the solar neutrino problem, and was anyway a feature of the paradigm that said what a Sun-like star ought to look like – until, that is, helioseismology tested the interior and demonstrated that a substantial fraction of the radiative interior is rotating at a rate no faster than the surface. This was a most unexpected result, again with implications not only for but also beyond solar and stellar physics (this time the domain of relativity theory).

This is an impressive list of successes. It might be tempting to think we can draw a line under various aspects of our descriptions of the Sun's interior. However, it turns out that for many we are a long way from being able to do so. To illustrate let us begin with what is, as I write, one of the hottest of these issues.

In 2004, Martin Asplund and his colleagues announced new estimates of the fractional abundances by mass of the heavy elements in the solar atmosphere. Getting a good determination is a very complicated business. Observations of the Fraunhofer lines formed by the heavy elements are used. But these observations must be combined with non-trivial models of how the lines are actually formed before an accurate estimate can be made. The analysis of Asplund and colleagues improved the models and the physical assumptions used to construct them.

The total abundance that was announced was important indeed, for it revised downwards the accepted value by just over forty per cent. This substantial change had implications for the standard solar model. Important ingredients the models require are the heavy-element abundances. We know the heavy elements leave their mark on the opacity in the interior, which in turn shows up readily in the sound-speed profile. When the new Asplund et al. value was adopted, it was found the model sound speeds changed by an amount that could be readily discriminated by the helioseismic data.

When the old heavy-element abundances had been used, the constructed models had given sound speeds that, throughout most of the interior, had fallen short by about a tenth of one per cent of those inferred from the best helioseismic data. There was a small increase in the difference just beneath the base of the convection zone, thought to be the signature of local mixing not present in the models. In the core the difference varied quite rapidly but remained at the level of a small fraction of one per cent. This core variation is believed to reflect the fact that the helium derived from the fusion reactions in the core does not simply stay put once made, as it does in the models.

When the new abundances were fed in, the model sound speeds dropped noticeably in the regions beneath the convection zone, where accurate determination of the opacities is vital. In the new picture that has emerged the model speeds are as much as one per cent smaller than the helioseismic values just below the base of the convection zone, and on average about a third of one per cent smaller throughout the radiative zone. The changes do not appear to penetrate deeply enough to be relevant to the solar neutrino problem (as has been shown in recent work by John Bahcall, Sarbani Basu, Marc Pinsonneault and Aldo Serenelli).

To a rough approximation, the new model profiles – with the revised abundances and gravitational settling and diffusion included – look not too dissimilar to the models of over a decade ago, which had the old abundances and no settling or diffusion. It might appear that the change to the heavy-element content of the models has approximately (and coincidentally) cancelled out the inclusion of settling and diffusion.[1] So, what is wrong with the models?

Historically, the sound-speed profiles of the models have undergone two fairly large adjustments (relative to the precision of the data). The first came from changes to the radiative opacities; the second from the inclusion of settling and diffusion effects. It is tempting to think the fresh discrepancies may point to the need to update the opacity calculations once more. Some revisions have recently been made, by members of an international consortium called the Opacity Project, which do go in the right direction. However, these alterations do not appear to be big enough to solve the problem. The differences may have something to do with mixing effects, but the origin remains far from clear. Only time will tell.

---

[1] The differences between the old model and helioseismic profiles tended to be larger throughout a greater portion of the radiative interior. Those between the new models and the currently best-available data fall off more rapidly deeper into the Sun.

Much remains to be done on the structure of the core itself. With the best helioseismic data to hand, scientists can make inversions to recover the sound-speed profile deep in the core. There the biggest differences between the observed profiles and models are currently around a quarter of one per cent. This sounds impressive – and it is. But if we are to believe the uncertainties (the error bars) on the sound-speed determinations, the differences are significant. The precision in the helioseismic data is so good that even the tiniest of discrepancies challenges our understanding of the processes taking place in the core.

But we must take care in accepting the current results at face value. There is a growing appreciation of the significance of subtle pitfalls that may lurk in the data analysis. Some relate to the influence of the Sun's activity cycle. The modes bear tiny scars from its effects, which originate in the near-surface layers where the likes of magnetic fields affect the frequencies. 'Scar' sounds like something harsh and unwanted – the effects are anything but when one uses the data to study the cycle itself. But the choice of word seems more appropriate if one wants to get a clean picture of the core and deep radiative interior. As dataset lengths increase, we are now reaching the point at which effects from the patchy nature of the activity on the surface may be influencing the picture of the core gleaned from the Sun-as-a-star data.[2]

We are still some way from being able to say anything definitive about how the rotation varies with depth in the core. Again, help will be at hand if the low-frequency p modes can be uncovered down to the fundamental tones. Although this should give real benefit for inferences made on the sound speed (from the frequencies), recent work suggests that we may begin to hit a brick wall regarding inference on the rotation from the splittings, and that really significant benefit may only accrue from the use of splittings of g modes. This supposes we have g mode

---

[2] This is to do with the fact that not all mode components are seen in the Sun-as-a-star data. It is therefore not possible to fully separate out contributions to the frequencies that come from spherically symmetric or non-symmetric effects. The patchy activity falls into the latter category.

data – which we do not. To my mind, the rotation problem alone is reason enough for us to redouble our efforts to uncover the g modes.

Co-ordinated efforts are ongoing. The Phoebus collaboration, led by Thierry Appourchaux, involves observers and theoreticians from many institutes who are making use of BiSON, GOLF, GONG, MDI and VIRGO data. The use of data from several instruments, data that have been collected at the same time, means the scientists can search for, and attempt to isolate, the common signal that is the signature of the g modes.

At the time of writing, we still await an unambiguous detection. One of the main results, I believe, to have come out of this work has been an appreciation that the going will remain tough if we continue to use the existing data in the same way. This statement says more than just that we need more data (which we very much do). We need to use the data in a more intelligent manner and also tailor the nature of the observations made by the instruments to better suit the demands of uncovering the g modes.

The ongoing searches are starting to push back boundaries in the use of data. Improvements to the observations will take more time, but again a great deal of thought and effort is being expended in this area. For example, several members of the GOLF consortium are making preparations to install a 'next generation' GOLF instrument – GOLF-NG – at Observatorio del Teide in Tenerife. This has been developed to get around the problem of the noise background from the convective granules bubbling away at the surface. The BiSON team are hard at work. And a French satellite, called *Picard*, is due to be launched soon which will offer a different approach again. It will exploit the fact the g mode signal is stronger when one's gaze is directed at the edge of the Sun's visible disc.

There are many reasons to be pessimistic about our ever detecting the g modes. I am rather more optimistic. When our desire to uncover them turns into a real *need* – and this will happen once we begin to hit the limits of what p modes can do for us – I believe we will naturally see more focus directed to this area. I cannot tell you the means by which we may finally find these modes. But I remain more confident than pessimistic that improvements to the observations, and advances in the data preparation and analysis – many of which remain unforeseen – will see us to this goal.

In Chapter 10 we saw that studies of the structure and dynamics of the outer layers of the solar interior are in good hands. The promise of what we can learn of magnetic structures and flows, and how they fit into the context of the solar activity cycle, is exciting indeed. The dynamo theorists will be seeking more input from helioseismology to pin down the mechanism that closes off the solar cycle – turning the east–west toroidal magnetic field back into the north–south poloidal field (the alpha and alpha-like effects). More detailed measurements of the properties of the tachocline will also be in the offing. A direct measurement, by helioseismology, of the magnetic field strength at the tachocline would be of great benefit to the dynamo modellers.

The local and global data needed to meet these aims will be provided in the future by both existing and up-coming observational programmes. GONG is now one of the permanent, flagship programmes of the National Solar Observatory and is currently in transition from what was a limited-lifetime project to a long-term one. The GONG instruments have been upgraded to observe the sun with higher resolution. This has shifted the emphasis from global to local seismology. When the GONG instrumentation was designed and its infrastructure put in place, components had not been chosen for their longevity. A systematic programme of upgrades is therefore being implemented to allow each site to meet future demands. The capability to return data in real time is being developed. Data analysis pipelines are also being put in place to return some of the exciting, new local-data products – like far-side images and flow maps – on a regular basis.

Plans call for the SOHO spacecraft to continue operations up to and just beyond the launch of the Solar Dynamics Observatory in 2008. The Solar Dynamics Observatory is one of the key elements of NASA's

Living With a Star programme and will carry a state-of-the-art helioseismology instrument called the Helioseismic Magnetic Imager. The Helioseismic Magnetic Imager is the new, improved, go-faster MDI and is being put together by a large team led by Phil Scherrer from Stanford. The key feature for the local seismologists is that the Helioseismic Magnetic Imager will not only have superior resolution to the MDI but will also have this all year round – the latter instrument operated at its peak capacity for only two months of every year.

The fact that high-resolution Doppler observations have been available simultaneously, from the MDI and GONG, has been of immense benefit to helioseismology. Only with independent data of this type has it been possible to confirm whether subtle effects really are solar in origin. As data quality improves and dataset lengths increase, more features like these will come to light. It is therefore vital that the simultaneous data capability be maintained. With the long-term future of GONG now secure, the Helioseismic Magnetic Imager and 'GONG++' will be in place to fulfil these needs as we move into the next decade.

ECHO continues its operations, as do TON and HiDHN on the high-degree side. On the Sun-as-a-star side, BiSON has been singled out by the UK's Particle Physics and Astronomy Research Council as one of five British-run observational programmes in solar–terrestrial physics that should receive long-term support. When SOHO – and therefore GOLF – ceases operations, this will leave BiSON as the only fully implemented Sun-as-a-star project. GONG and the Helioseismic Magnetic Imager will have a low-degree capability; the hope is that the ground-based trial observations of GOLF-NG will lead to it being selected for a future space mission.

That is a snapshot of where helioseismology stands at present. Let us finish by looking further afield – to the stars. When George Isaak and his colleagues uncovered the global nature of the solar p modes, the starting gun fired for efforts to find Sun-like, low-degree, core-penetrating modes in other stars. Both George's and Eric's teams had, after all, adopted a 'Sun-as-a-star' approach to the observations.

Attempts to observe solar-like oscillations – by which we mean small-amplitude pulsations excited by convection in the near-surface layers of a star – took over a decade to bear fruit. The technical challenges to overcome were immense. The difficulty for any keen observer is the paltry light level of even the brightest of stars compared with the Sun. This makes it hard to spot the tiny observable effects the modes give rise to. Since the Doppler velocity technique is far less susceptible to the effects of the intervening terrestrial atmosphere, once improvements had been made to overcome the handicap of relative brightness, it was from Doppler data that *asteroseismology* of Sun-like stars became a reality.

The initial objective for the stellar observers was to demonstrate the presence of an excess bump of power, significantly over and above the adjacent background, in a frequency spectrum of the observations. The excess power had to lie within a plausible range of frequency. To really convince the community, what was needed was evidence of a near-regular picket fence of peaks, the signature of the overtone structure expected of low-degree modes. The breakthrough came in the mid 1990s.[3] With the benefit of hindsight it now appears the first bone fide report of solar-like oscillations that had an accompanying estimate of the picket-fence spacing came from observations of the star Eta Bootes by Hans Kjeldsen, Tim Bedding, Michael Viskum and Soren Frandsen.[4] This star has completed its Main Sequence lifetime and is therefore in a more evolved state than the Sun.

Observations of oscillations on Alpha Centauri A, reported by Jesper Schou and Derek Buzasi, represented the first detection on a star

---

[3] For an in-depth review of the field, and important references, see T.R. Bedding and H. Kjeldsen, *Publications of the Astronomical Society of Australia*, 20, 2003, p. 203.

[4] The first report of a significant excess of power attributable to oscillations may be that of Tim Brown et al.'s observations of the star Procyon. T.M. Brown, R.L. Gilliland, R.W. Noyes and L.W. Ramsey, *Astrophysical Journal*, 368, 1991, p. 599.

closely resembling the Sun. This evidence came from data collected in 1999 by the star camera on a NASA satellite called the Wide-Field Infrared Explorer. Because it was above the atmosphere, the camera – although not designed for the purpose – was able to track the tiny changes in intensity given by the oscillations. Modes on the same star were uncovered even more dramatically from ground-based Doppler observations by François Bouchy and Fabien Carrier in 2001.

Detections have now been made on more than a dozen stars. Ground-based observations are being spearheaded by Tim Bedding, Paul Butler, Hans Kjeldsen, and François Bouchy and Fabien Carrier. These observations use the Doppler technique. The Canadian Microvariability and Oscillation of Stars (MOST) satellite was launched in 2004. It is the first of the bespoke intensity missions to fly. Jaymie Matthews and his team will be using MOST to look at a selection of stellar targets. The French Convection Rotation and Planetary Transits (COROT) spacecraft is due to be launched in 2006. It has a detailed programme of observations planned, including monitoring some stars for periods up to three months.

Asteroseismology is on the brink of an explosion in the amount of data available to analyse. Seismic observations of other stars offer the prospect of testing stellar evolution theory using many stars, rather than just one (the Sun). A range of targets are potentially on offer that are at different points along their evolutionary life cycles. The theories will therefore be subjected to exacting examination – how well will they be able to describe the interiors of such a cross-section of stars?

For the foreseeable future the data will be limited to the low-degree modes. The smaller number of modes available, compared with data on the Sun, presents a major challenge to scientists wishing to uncover the interior structures of stars. The signs are, however, promising. Techniques to use, and invert, low-degree-only data are being developed.

Inference about the activity behaviour of other stars will also be possible. Since solar-cycle effects can be uncovered in Sun-as-a-star data,

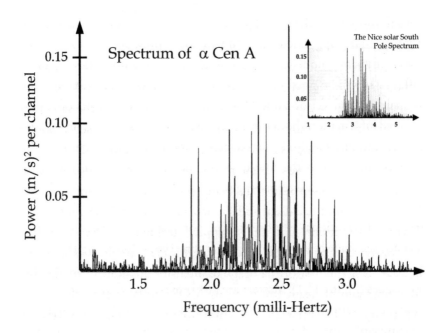

Figure 11.1. Observed power spectrum of the solar-like star Alpha Centauri A (α Cen A), from data collected in 2001 by Paul Butler, Tim Bedding, Hans Kjeldsen and colleagues. The inset shows, for comparison, the Nice South Pole solar spectrum made from data collected over the New Year period in 1979. Stellar data reproduced with the kind permission of Tim Bedding and Hans Kjeldsen, from R.P. Butler et al., *Astrophysical Journal*, 600, 2004, p. L75; South Pole spectrum from *Nature* (see Chapter 5), reproduced with the kind permission of the Nature Publishing Group.

it follows that similar-sized variations will also be apparent from long-term monitoring of stellar targets. Observations of activity on other stars – from the sizes of emissions produced by certain atomic lines – already suggest that some may be in a Maunder-Minimum-like state. It will be interesting to see what the seismic activity distribution looks like once long-term monitoring of a selection of asteroseismic targets is established.

Stars picked for observation that end up showing little, if any, activity should yield very clean inference about their interior structure. The contribution of patchy solar-like activity in the near-surface layers will not have to be suppressed – since it will not be present.

The dynamo modellers will await the asteroseismic data with great interest. It will be possible to begin to probe the characteristics of the sub-surface convection zones, beginning with measurements of depth. Limited information about the distributions of active regions on the surface of stars should also be extractable from the frequency shifts of the components seen in the low-degree data. A good deal of care will, though, be needed in interpreting such data.

In 2005 observational asteroseismology of Sun-like stars has just about reached the stage helioseismology was at in 1980 – following the publication of the South Pole Sun-as-a-star spectrum of Eric Fossat and his colleagues (Figure 11.1). Matters are rather more evolved on the analysis and interpretation side. Asteroseismology has the benefit of a quarter of a century of knowledge, about techniques and their uses, accumulated by helioseismology.

Meanwhile helioseismology has entered a new phase, with important commitments made to continue long-term monitoring of the solar oscillations, and new measurements and science becoming possible. It will be fascinating to see where helio- and asteroseismology have taken us in a further quarter-century. Watch this space.

INDEX

Index

Index

Index

Index

Index

Index